Advances in
Ceramic Armor III

Advances in Ceramic Armor III

A Collection of Papers Presented at the 31st International Conference on Advanced Ceramics and Composites January 21–26, 2007 Daytona Beach, Florida

Editor

Lisa Prokurat Franks

Volume Editors

Jonathan Salem

Dongming Zhu

The American Ceramic Society

BICENTENNIAL

1807

WILEY

2007

BICENTENNIAL

WILEY-INTERSCIENCE

A John Wiley & Sons, Inc., Publication

Published by John Wiley & Sons, Inc., Hoboken, New Jersey.
Published simultaneously in Canada.

For general information on our other products and services or for technical support, please contact our Customer Care Department within the United States at (800) 762-2974, outside the United States at (317) 572-3993 or fax (317) 572-4002.

Wiley also publishes its books in a variety of electronic formats. Some content that appears in print may not be available in electronic format. For information about Wiley products, visit our web site at www.wiley.com.

Wiley Bicentennial Logo: Richard J. Pacifico

Library of Congress Cataloging-in-Publication Data is available.

ISBN 978-0-470-19636-6

10 9 8 7 6 5 4 3 2 1

Contents

Preface

The 31st International Conference and Exposition on Advanced Ceramics & Composites, January 22-26, 2007, in Daytona Beach, Florida marks the fifth consecutive year for modelers, experimentalists, processors, testers, fabricators, manufacturers, managers, and simply ceramists to present and discuss the latest issues in ceramic armor. This annual meeting is organized and sponsored by The American Ceramic Society (ACerS) and the ACerS Engineering Ceramics Division, and is affectionately known as the Cocoa Beach conference because it has been held in Cocoa Beach, Florida since its inception. Although there was concern about the risk of a change in location of such a time honored tradition as the Cocoa Beach conference, the move up the coast to Daytona Beach proved successful for the Ceramic Armor Symposium. Currently, there is no other unclassified forum in government, industry, or academia that brings together these interdependent disciplines to address the challenges in ceramic armor, and consequently, we needed a more spacious venue.

For each of the past five years, we have been privileged to have technical experts who are not only enthusiastic about their work, but also willing to address the tough or controversial aspects in lively discussion. This meeting continued the tradition of an unclassified forum on an unprecedented international scale with representatives from the U.K., Germany, France, Italy, Sweden, Japan, Brazil, China, and Thailand.

The sessions included presentations from Patrik Lundberg of the Swedish Defence Research Agency, "Research in Sweden on Dwell in Ceramics"; Daniel C. Harris of the Naval Air Systems Command, "Supersonic Water Drop Impact Resistance of Transparent Ceramics"; Carl F. Cline of Advanced Materials Technology Int. and previously the Lawrence Radiation Laboratory, University of California, Livermore, "Light Weight Armor: 40 Years Later"; and Victor A. Greenhut of Materials Engineering at Rutgers University, "Severe Inclusions Preferred Fragmentation Path in Silicon Carbide Armor Ceramics." For 2007, topical areas solicited for abstracts were expanded to six and included transparent ceramics, predicting ceramic armor performance, damage characterization, novel material concepts, protection against fragments, and manufacturing challenges. The papers in this proceedings have been grouped into the following categories: General/Overview; Glasses and Transparent Ceramics; Opaque Ceramics; and Damage and Testing.

Although special thanks must go to the Organizing Committee for its exceptional efforts to support the Ceramic Armor Symposium in a year where funding and travel approvals threatened to de-evolve completely, I want to especially extend my sincere thanks to the participants. With your reasonable and concise feedback we have been able to keep the successes and make improvements where needed. Since the first Ceramic Armor session, many of you have ensured that the "Cocoa Beach" conference is your must-attend-event, and the core group of participants has continued to grow. Five years ago when we had to ask that available seating double after the first talk as well as ask the first speaker to give his presentation again at the end of the day to accommodate participants that couldn't make it into the room or hear from the hallway, success seemed likely to be associated with such a novel topic or possibly a kind of beginners' luck. The Organizing Committee could not have predicted that 2007 would again be a standing room only, spill over into the hallway kind of year; there were four times as many seats as five years ago and nearly double that available last year! Thank you ladies and gentlemen. The Organizing Committee has its work cut out for it in 2008, and we accept the challenge.

LISA PROKURAT FRANKS
U.S. Army TARDEC

Introduction

2007 represented another year of growth for the International Conference on Advanced Ceramics and Composites, held in Daytona Beach, Florida on January 21-26, 2007 and organized by the Engineering Ceramics Division (ECD) in conjunction with the Electronics Division (ED) of The American Ceramic Society (ACerS). This continued growth clearly demonstrates the meetings leadership role as a forum for dissemination and collaboration regarding ceramic materials. 2007 was also the first year that the meeting venue changed from Cocoa Beach, where it was originally held in 1977, to Daytona Beach so that more attendees and exhibitors could be accommodated. Although the thought of changing the venue created considerable angst for many regular attendees, the change was a great success with 1252 attendees from 42 countries. The leadership role in the venue change was played by Edgar Lara-Curzio and the ECD's Executive Committee, and the membership is indebted for their effort in establishing an excellent venue.

The 31st International Conference on Advanced Ceramics and Composites meeting hosted 740 presentations on topics ranging from ceramic nanomaterials to structural reliability of ceramic components, demonstrating the linkage between materials science developments at the atomic level and macro level structural applications. The conference was organized into the following symposia and focused sessions:

- Processing, Properties and Performance of Engineering Ceramics and Composites
- Advanced Ceramic Coatings for Structural, Environmental and Functional Applications
- Solid Oxide Fuel Cells (SOFC): Materials, Science and Technology
- Ceramic Armor
- Bioceramics and Biocomposites
- Thermoelectric Materials for Power Conversion Applications
- Nanostructured Materials and Nanotechnology: Development and Applications
- Advanced Processing and Manufacturing Technologies for Structural and Multifunctional Materials and Systems (APMT)

- Porous Ceramics: Novel Developments and Applications
- Advanced Dielectric, Piezoelectric and Ferroelectric Materials
- Transparent Electronic Ceramics
- Electroceramic Materials for Sensors
- Geopolymers

The papers that were submitted and accepted from the meeting after a peer review process were organized into 8 issues of the 2007 Ceramic Engineering & Science Proceedings (CESP); Volume 28, Issues 2-9, 2007 as outlined below:

- Mechanical Properties and Performance of Engineering Ceramics and Composites III, CESP Volume 28, Issue 2
- Advanced Ceramic Coatings and Interfaces II, CESP, Volume 28, Issue 3
- Advances in Solid Oxide Fuel Cells III, CESP, Volume 28, Issue 4
- Advances in Ceramic Armor III, CESP, Volume 28, Issue 5
- Nanostructured Materials and Nanotechnology, CESP, Volume 28, Issue 6
- Advanced Processing and Manufacturing Technologies for Structural and Multifunctional Materials, CESP, Volume 28, Issue 7
- Advances in Electronic Ceramics, CESP, Volume 28, Issue 8
- Developments in Porous, Biological and Geopolymer Ceramics, CESP, Volume 28, Issue 9

The organization of the Daytona Beach meeting and the publication of these proceedings were possible thanks to the professional staff of The American Ceramic Society and the tireless dedication of many Engineering Ceramics Division and Electronics Division members. We would especially like to express our sincere thanks to the symposia organizers, session chairs, presenters and conference attendees, for their efforts and enthusiastic participation in the vibrant and cutting-edge conference.

ACerS and the ECD invite you to attend the 32nd International Conference on Advanced Ceramics and Composites (http://www.ceramics.org/meetings/daytona2008) January 27 - February 1, 2008 in Daytona Beach, Florida.

JONATHAN SALEM AND DONGMING ZHU, Volume Editors
NASA Glenn Research Center
Cleveland, Ohio

General/Overview

RESEARCH IN SWEDEN ON DWELL IN CERAMICS

Patrik Lundberg
FOI, Swedish Defence Research Agency, Defence & Security Systems and Technology
SE-14725 Tumba, Sweden

ABSTRACT

The first documented high velocity tungsten long rod experiment in Sweden that clearly demonstrated the occurrence of dwell was conducted in 1987. In a set of impact experiments with different types of confined ceramic targets, one of the tests with a titanium diboride target showed a tremendous protection capability compared to the other targets tested. Post-mortem examinations showed that in this target, the tungsten rod had initially penetrated the ceramic in an ordinary way to a depth of some mm but then started to flow radially instead of continuing to penetrate through the sample. This single result was not possible to repeat and it was handled as an anomaly since the dwell phenomenon was not established at that time, and the experiment was soon forgotten. Ten years and over 200 dwell related experiments later, it is now well established that high velocity dwell and interface defeat is a result of the exceptional strength of the ceramics, and that this high strength is available if target damage can be suppressed. The major parts of the basic research in Sweden on dwell and interface defeat has been performed at FOI and focus has been on the loading conditions on the ceramic front face. Both stationary and non-stationary loading conditions relevant for an armour implementation have been analysed theoretically and studied experimentally in order to determine the performance limits of different ceramic materials.

INTRODUCTION

Initial studies on hard materials for armour applications started in Sweden during the fifties. At this time, geological materials like granite and diabas were tested for armour applications, primarily for protection against shape charge warheads. The first report on ceramics for protection against penetrating threats came in the late sixties and the first experiment with ceramic armour for protection against projectiles was made in 1973. The armour material at that time was an alumina used for electrical isolation which was produced in Sweden by IFÖ keramik AB.

During the early eighties, a special test facility with a smooth bored gun (calibre 30 mm) was set-up at FOA (later FOI) in order to facilitate systematic studies of the interaction between small scale long rod projectiles and various types of armours. An initial study of the protection capability of high quality ceramics was initiated 1986. The materials (Al_2O_3, Si_3N_4, B_4C, SiC and TiB_2) were produced by ASEA Cerama AB using a glass-encapsulated hot iso-static pressing technique to produce large samples of full dense materials.

The first documented high velocity tungsten long rod experiment that clearly demonstrated the occurrence of dwell was conducted in 1987. In a set of impact experiments at velocities of around 1550 m/s with different types of confined ceramic targets, one of the tests with a titanium diboride target showed a tremendous protection capability compared to the other targets tested. Post-mortem examinations showed that in this target, the tungsten rod had initially penetrated the ceramic in an ordinary way to a depth of some mm but then started to flow radially instead of continuing to penetrate through the sample. This single result was not possible to repeat and it was handled as an anomaly since the dwell phenomenon was not established at that time, and the experiment was soon forgotten. Figure 1 shows the recovered ceramic from this specific test.

FIGURE 1. Post mortem examination of the titanium diboride "dwell target". The impact direction is from the top.

A ceramic armour consortium was formed by the Swedish Defence, Swedish ceramic industries and Swedish armour producers in 1990. The overall objective was to develop know-how for design and manufacture of weight and cost effective add-on ceramic armour for Swedish Military vehicles and vessels, especially for the CV-90 IFV. The program included basic as well as applied studies but, despite a large experimental program on the interaction between small and medium calibre projectiles and various types of ceramics lasting nearly four years, only one test indicating dwell was reported (a small scale experiment with titanium diboride).

The fist time dwell was investigated in an impact experiment in Sweden was in 1996. At this time information about the work by Hauver *et al*. had just reached Sweden and the first small scale reverse impact experiments with boron carbide targets were just to be conducted. Several of these experiments showed dwell to occur for considerable periods of time and although the initial idea of the experiments was not primarily to study dwell, these experiments opened for a Swedish research program on dwell and interface defeat. This effort is still continuing, but now with a smaller founding.

Ten years and over 200 dwell related experiments later, it is well established that high velocity dwell and interface defeat is a result of the exceptional shear strength of the ceramics, and that this high strength is available if target damage can be suppressed. The major parts of the basic research in Sweden on dwell and interface defeat has been performed at FOI (earlier FOA) and focus has been on the loading conditions on the ceramic front face. Both stationary and non-stationary loading conditions relevant for an armour implementation have been analysed theoretically and studied experimentally in order to determine the performance limits of different ceramic materials.

This paper summarises the work done at FOI on dwell and interface defeat for high velocity long rod projectiles impacting generic ceramic targets.

DWELL AND INTERFACE DEFEAT

Dwell and interface defeat was systematically studied and first reported by Hauver *el al.*[1,2] and later by Rapacki *et al.*[3]. They showed that by using different devices for load distribution and attenuation, including shrink-fitted or hot-pressed confinements, target damage could be suppressed. In this way it was possible to design thick ceramic armour systems capable of defeating long-rod projectiles at high impact velocities on the surface of the ceramic. This capability, named dwell or, if the threat was completely defeated, interface defeat, signifies that the projectile material is forced to flow radially outwards on the surface of the ceramic material without penetrating significantly. The mode of interaction is illustrated in Fig. 2 schematically as well as by a low velocity water jet and a flash X-ray picture from a high velocity impact experiment.

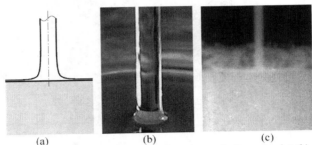

(a)   (b)   (c)

FIGURE 2. Dwell illustrated schematically (a), for a low velocity water jet (b) and for a high velocity tungsten long rod projectile (c) (flash X-ray picture).

The aim of the research at FOI was to estimate the transition velocity (i.e., the maximum velocity at which interface defeat can be maintained) at various loading conditions for different combinations of projectiles and ceramic targets. Another goal was to identify critical velocities which may serve as upper and lower bounds to the transition velocity, especially those related to the material properties of the projectile and the target.

In [4-7], the transition velocity was studied for different types of ceramic materials (boron carbide, silicon carbide, titanium diboride and Syndie) and in [8] it was estimated for one type of ceramic material (four grades of silicon carbide). The possibility to maintain dwell for a long period of time was studied in [9].

The influence of the properties of the projectile on the transition velocity was examined in [7] and [10]. In [7], use was made of two different projectile materials, a tungsten heavy alloy and molybdenum, with large difference in density. In [10] the transition velocity was estimated for a conical projectile and compared to that of a cylindrical projectile. In [11] the transition velocity of an unconfined target was studied.

In [7] the influence of the strength of the ceramic material on the transition velocity was studied using two models, viz., one for the surface pressure during interface defeat and one for the yield conditions in the ceramic material. The contact load during dwell at normal impact was further analysed in [12] and [13].

The constitutive models JH1 and JH2 by Johnson and Holmquist[14,15] were used to model the response of the ceramic material. JH1 was used in [10,12] under assumed conditions of interface defeat and JH2 was used in [5] to model both interface defeat and penetration.

EXPERIMENTAL TECHNIQUES

A two-stage light-gas gun was used for the impact experiments described in [4-9,11]. The pump-tube diameter of the gun is 80 mm and the diameter of the launch-tube is 30 mm. The impact experiments are performed in a tank connected directly to the launch tube. The experiments in [10] were performed with a 30 mm powder gun which has a similar impact tank and instrumentation as that of the light-gas gun. Figure 3 shows part of the light-gas gun facility.

FIGURE 3. High pressure section, launch tube, impact tank and flash X-ray system of the two stage light-gas gun.

All the impact experiments were performed in a small scale with simple projectile and target geometries. Either the projectile was launched against a stationary target (direct impact) or the reverse impact technique was used. The latter means that the target is launched against the projectile, which is fixed in front of the muzzle of the launch tube. These two impact techniques are kinematically equivalent. Examples of projectile set-ups in reverse impact experiments are shown in Fig. 4.

FIGURE 4. Examples of projectile set ups in reverse impact tests.

Several X-ray flashes (150 kV - 450 kV) were used to depict the penetration process. The flash X-ray technique makes it possible to see through the target and confinement materials. Therefore, it offered the possibility to study the penetration process in detail. An example of the X-ray set-up is shown in Fig. 5.

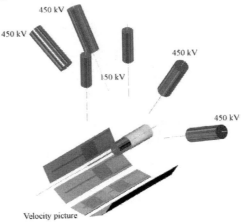

FIGURE 5. Typical X-ray set-up.

PROJECTILE AND TARGET MATERIALS USED

Two tungsten heavy alloys, DX2-HCMF from Pechiney and Y925 from Kennametal Hertel AG, were used as projectile materials. The ceramics used were produced by pressure assisted sintering (SiC-B, SiC-N, SiC-SC-1RN, SiC-HPN) and hot iso-static pressing (B$_4$C-AC, SiC-AC, TiB$_2$-AC). The diamond material in [7], Syndie, is a polycrystalline diamond composite produced by De Beers. In the reverse impact experiments, the projectile was mounted in front of the muzzle of the launch tube using either thin threads or a fixture of Divinycell. Two types of steel were used as confinement. Initially a tempered steel, SIS-2541-03, was used. This is comparable to AISI/SAE 4340. It was later replaced by a MaraModelling steel (Mar 350). The front cover was initially made of steel (2541-03), but later pure copper was used.

MODELLING AND ANALYSES

In [10], critical velocities related to the transition between interface defeat and penetration were determined for cylindrical and conical projectiles. Idealised loading conditions were used. Thus, the transient part of the loading and the influence of the radial growth of the surface load for the conical projectiles were not taken into account. The loading conditions are illustrated in Fig. 6.

The surface pressures and the corresponding critical velocities at which plastic yield is initiated at a point on the axis below the surface and the domain of plastic yield reaches the target surface were established in [7]. These states of yield, labelled incipient plastic yield (IY) and full plastic yield (FY), respectively, are illustrated schematically in Fig. 7.

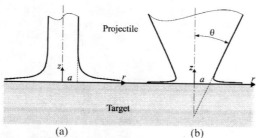

(a)   (b)

FIGURE 6. Flow of (a) a cylindrical and (b) a conical projectile during interface defeat.

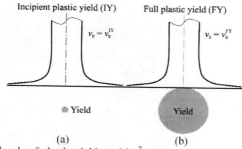

(a)   (b)

FIGURE 7. Critical levels of plastic yield used in [7]. (a) Incipient plastic yield (IY) and (b) full plastic yield (FY).

For a projectile with bulk modulus K_p, yield strength σ_{Yp} and density ρ_p, an approximate relation for the maximum surface pressure on the axis of symmetry was obtained as

$$p_0 \approx q_p\left(1+\frac{1}{2\alpha}+3.27\beta\right),$$

(1)

where $\alpha = K_p/q_p$, $\beta = \sigma_{Yp}/q_p$ and $q_p = \rho_p v_0^2/2$. Here v_0 is the impact velocity of the projectile and q_p is the stagnation pressure of an ideal fluid with density ρ_p and velocity v_0. The dimensionless parameters $\alpha \gg 1$ and $\beta \ll 1$ relate elastic and plastic effects, respectively, to the effect of inertia. The modelling of the surface pressure was refined and expanded in [13], where the result (1) was confirmed.

The radial distribution of the surface pressure was approximated by a pressure profile determined experimentally for a low-velocity water jet. The assumed pressure distribution in combination with Boussinesq's elastic stress field solution gives the relation $p_0 = (2.601+2.056\nu)\tau_0$ between the maximum surface pressure p_0 and the maximum shear stress τ_0 in the ceramic. An approximate critical surface pressure corresponding to incipient yield (IY), was obtained by putting the maximum shear stress τ_0 equal to the shear yield strength τ_{Yc}.

A plastic slip-line solution for the indentation of a rigid punch into a rigid-plastic half-space was used to obtain an approximate critical surface pressure corresponding to full plastic yield (FY). According to this solution, the relation $p_{mean} = 5.70\tau_{Yc}$ between the mean pressure and the shear yield strength τ_{Yc} is valid. Here the mean pressure was taken as the maximum surface pressure, i.e. $p_0 = 5.70\tau_{Yc}$.

These critical surface pressures corresponding to incipient yield (IY) and full plastic yield (FY), give the approximate critical surface pressure interval

$$(1.30 + 1.03\nu)\sigma_{Yc} \le p_0 \le 2.85\sigma_{Yc},\tag{2}$$

where $\sigma_{Yc} = 2\tau_{Yc}$ is the yield strength in uniaxial compression according to Tresca's hypothesis and ν is the Poisson's ratio of the ceramic. As relation (1) provides a relation between the maximum surface pressure p_0 and the impact velocity v_0, this relation and inequalities (2) give a critical velocity interval for the impact velocity v_0.

In [10], both cylindrical and conical projectile loading was studied by means of numerical modelling. The surface pressures and the corresponding critical velocities were established at which damage is initiated at a point on the axis below the surface, the domain of damage reaches the target surface, and a ring-shaped surface failure is initiated (damage and failure defined according to the Johnson and Holmquist constitutive models JH1 and JH2[14,15]).

These states of damage, labelled incipient damage (ID), full damage (FD) and surface failure (SF), are illustrated schematically in Fig. 8.

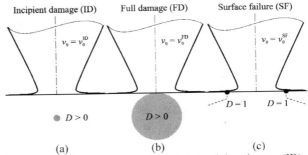

FIGURE 8. Critical levels of damage and failure used. (a) Incipient damage (ID), (b) full damage (FD) and (c) surface failure (SF). Reprinted from Reference [10] with permission from Elsevier.

The code Autodyn was used to determine the surface pressure on the target and the resulting target damage due to impact by conical and cylindrical projectiles. The simulations, two-dimensional with rotational symmetry, were carried out in two steps. In the first step, the pressure distribution on a flat, rigid and friction-free target surface was determined (Eulerian simulations, Johnson and Cook constitutive model [16]). In the second step, the surface pressure distribution was applied to a deformable target, and the damage of the target material was assessed (Lagrangian simulation, Johnson and Holmquist constitutive model JH1). Since each critical level of damage and failure corresponds to a critical surface pressure, it was possible to determine the critical impact velocities v_0^{ID}, v_0^{FD} and v_0^{SF}.

TRANSITION VELOCITY VERSUS MATERIAL PROPERTIES AND CONFINEMENT

For specific studies of the transition velocity, two slightly different targets were developed (designated Target A and B). For Target A, used in [7], the threaded front and back steel plugs were locked axially by rings which were electron-beam welded onto the tube. This was done in order to avoid early confinement failure. This target was used for estimations of the transition velocities for two grades of silicon carbide, titanium diboride and Syndie.

In order to facilitate the comparison of the transition velocity of materials with only small differences in mechanical properties, the target configuration was further developed. In particular, the front cover and the confinement tube were redesigned so that the transition velocity could be assessed in a more reproducible way. The increased simplicity of the target also facilitated modelling of dwell.

In this case, the confinement consisted of a thick-walled steel cup with a circular copper cover. The copper cover was expanded in its central part to a cylinder with diameter 4 mm and length 8 mm. It was glued onto the steel tube along the rim. This target (Target B) was used in [8] to determine transition velocities for four grades of silicon carbide, SiC-B, SiC-N, SiC-SC-1RN and SiC-HPN.

In order to study the influence of confinement on the transition velocity for SiC-B, a third unconfined target was used [11]. The copper cover was similar to the one used for Target B.

The geometries of Targets A, B and C are shown in Fig. 9.

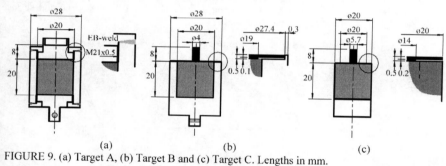

(a)   (b)   (c)

FIGURE 9. (a) Target A, (b) Target B and (c) Target C. Lengths in mm.

Yield strength of the target material

The transition velocity was determined for different types of ceramic materials (SiC-B, SiC-AC, TiB$_2$-AC, and Syndie) with relatively large variation in Vickers hardness. The Vickers hardness was used to estimate the yield strengths of the ceramics. Figure 10 shows the transition velocities determined from the impact experiments together with critical velocities corresponding to incipient yield (IY) and full plastic yield (FY). The critical velocities were determined from equation (1) and inequalities (2). The lower curve corresponds to Poisson's ratio $v = 0.07$.

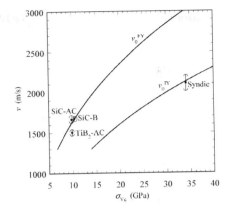

FIGURE 10. Transition velocity v_0^* from impact tests and critical velocities v_0^{IY}, v_0^{FY} versus target yield strength σ_{Yc} (solid curves). The ends of the error bars represent the highest velocity without penetration ($\cup$) and the lowest velocity with penetration ($\cap$), respectively. Reprinted from Reference [7] with permission from Elsevier.

Fracture toughness of the target material

The transition velocity was determined for four grades of silicon carbide having relatively small variations in hardness and toughness. The transition velocities are plotted versus normalised hardness and toughness in Fig. 11.

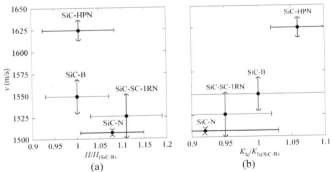

FIGURE 11. Transition velocity v_0^* versus (a) Vickers hardness H and (b) fracture toughness K_{Ic}. The lower and upper limits correspond to the highest velocity without penetration ($\cup$) and the lowest velocity with penetration ($\cap$), respectively. The left and right limits correspond to the standard deviation in hardness and fracture toughness. Reprinted from Reference [8] with permission from Elsevier.

Influence of the confinement

The transition velocity for one grade of silicon carbide (SiC-B) was determined for three different target designs representing different confinement conditions. The estimated transition velocities v_0^* are indicated as shaded areas in Fig. 12. The dashed curve in the figure is based on Tate's model.

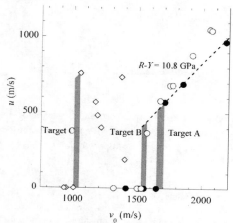

FIGURE 12. Penetration velocity u versus impact velocity v_0 in SiC-B for Target A (●), Target B (○) and Target C (◇). The shaded areas indicate transition regions. The dashed curve is based on Tate's model.

Density of the projectile material

Target A in Fig. 9 was used to study the influence of material properties of the projectile on the transition velocity. Two different projectile materials, a tungsten heavy alloy and molybdenum, with densities 17600 kg/m^3 and 10220 kg/m^3, respectively, were used. The observed intervals for the transition velocities are shown in Fig. 13 together with the penetration velocities estimated from Tate's model.

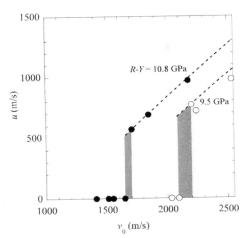

FIGURE 13. Penetration velocity u versus impact velocity v_0 for DX2-HCMF ($\bullet$) and Mo ($\circ$) impacting SiC-B. The shaded areas indicate transition regions. The dashed curves are based on Tate's model.

TRANSITION VELOCITY VERSUS PROJECTILE GEOMETRY

In order to assess the influence of a conical nose shape on the transition velocity, experimental tests with conical and cylindrical WHA projectiles impacting silicon carbide targets were performed together with numerical simulations using the Autodyn code.

As pointed out above, the simulations were run in two steps. In the first, the surface pressures on the target were determined for different impact velocities under assumed condition of interface defeat. In the second, these surface pressures were applied to the target in order to obtain critical states of damage and failure related to the transition between interface defeat and penetration, see Fig. 8.

In the experiments, projectiles with cylindrical and conical fronts of nominal length 30 mm, fabricated by grinding sintered Y925 rods, were used. The conical front projectiles had apex angle 10° (half apex angle $\theta = 5°$).

The target consisted of a silicon carbide cylinder (SiC-B) shrink-fitted into a cylindrical steel cup. In the tests with the cylindrical projectiles and the conical projectiles with the larger front diameter, a small cylindrical copper cover was glued along its rim to the surface of the target. The geometries of the projectiles, the target and the copper cover are shown in Fig. 14.

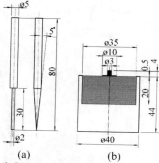

(a)   (b)

FIGURE 14. (a) Projectiles and (b) target with copper cover. Lengths in mm. Reprinted from Reference [10] with permission from Elsevier.

The dependence on half apex angle θ of the experimentally estimated transition velocity v_0^* and the calculated critical impact velocities v_0^{ID}, v_0^{FD} and v_0^{SF}, corresponding to incipient damage, full damage and surface failure, respectively, are shown in Fig. 15.

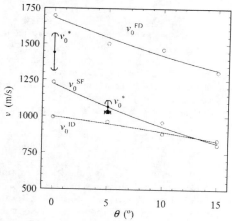

FIGURE 15. Transition velocity v_0^* from impact tests and critical velocities v_0^{ID}, v_0^{FD}, v_0^{SF} from simulations versus half apex angle θ. The ends of the error bars represent the highest velocity without penetration ($\cup$) and the lowest velocity with penetration ($\cap$). The solid curves are second-degree polynomial fits to the simulations ($\circ$). Reprinted from Reference [10] with permission from Elsevier.

DISCUSSION

In order to establish nearly quasi-static loading experimentally an attenuating device has to be used. In the FOI studies, this device was the front plug, initially made of steel, with a thickness sufficient enough to considerably reduce the initial transient load on the surface of the ceramic. This attenuation device was later changed to a copper cover that established erosion of the projectile before it reached the ceramic surface. During this process and the ensuing radial flow of the projectile material, the copper cover was eroded so that the projectile material was left unsupported on the target surface. The results in [8,9] show that after radial flow has been established, it continues steadily for a long time (several hundreds of micro seconds in [9]).

In [7], the surface pressures and the corresponding critical velocities were established at which plastic yield is initiated at a point on the axis below the surface and the domain of plastic yield reaches the target surface. These states of yield are shown as lower and upper critical velocities in Fig. 10. It can be noted that the estimated transition velocities in the experiments are close to these critical velocities.

In Fig. 11, SiC-N shows the lowest transition velocity, slightly above 1500 m/s, and SiC-HPN the highest, slightly above 1600 m/s. As the target configurations and impact conditions were the same in these tests, the difference between these transition velocities should depend on the mechanical properties of the ceramic materials. Furthermore, as the maximum surface pressure p_0 produced by the projectile during interface defeat is approximately proportional to the square of the impact velocity v_0 according to equation (1), i.e. $p_0 \propto v_0^2$, the 8% higher transition velocity for SiC-HPN than for SiC-N corresponds to a 16% higher maximum surface pressure. If the transition velocity v_0^* and the corresponding maximum surface pressure p_0^* are related to the shear yield strength τ_{Yc} of the ceramic material through $\tau_{Yc} \propto p_0^* \propto v_0^{*2}$ as proposed in [7], the strength of SiC-HPN should be about 16% higher than that of SiC-N. As can be seen in Fig. 11(a), however, the ranking in terms of hardness of the ceramic materials is not the same as that in terms of transition velocity. Actually, the hardness of SiC-HPN is 7% lower than that of SiC-N. This indicates that hardness alone, which reflects the shear yield strength, is not sufficient for estimating of the relative performance of ceramic materials in terms of transition velocities. Thus, the transition velocity cannot be controlled by plastic flow alone.

Figure 11(b) shows the transition velocity versus fracture toughness. Here, an increase in transition velocity corresponds to an increase in fracture toughness, although the uncertainty in fracture toughness is large. The maximum surface pressure is 16% and the fracture toughness 15% higher for SiC-HPN than for SiC-N. This observation indicates that, under the prevailing conditions, fracture may have more influence than plastic flow on the transition from interface defeat to penetration. As a consequence, the observed transition velocities may not be the maximum ones achievable. By suppressing of the initiation and propagation of cracks through increase of the confining pressure, it may be possible to increase the transition velocities.

The influence of confining pressure may be one reason for the difference in transition velocity between the three SiC-B targets used in Fig. 12. The target designs are shown in Fig. 9. Target A, with a front cover (and back plug) welded to the confining tube, gives 8% higher transition velocity as compared to Target B and 63% higher velocity as compared to Target C.

The influence of the material properties of the projectile on the transition and penetration velocities is shown in Fig. 13. The transition velocity intervals determined for a Tungsten projectile, DX2-HCMF, and a Mo projectile impacting SiC-B are indicated by grey zones. The assumption that the transition from interface defeat to penetration occurs at a critical surface pressure gives $p^*_{Mo} = p^*_{WHA}$. Using of equation (1) and material data for DX2-HCMF and Mo then gives the relation between the transition velocities as $v^*_{Mo} \ v^*_{WHA} \approx 1.34$. This agrees reasonably well with the experimental result $v^*_{Mo} \ v^*_{WHA} \approx 1.28$.

In the tests with conical projectiles, the size of the copper cover was large compared to the initial diameter of the projectile tip. This means that the initial loading of the target surface was smoother than in the tests with cylindrical projectiles. Yet, the transition velocity for conical projectiles and copper-covered targets was significantly lower than that for the cylindrical projectiles for the same kind of target. The decrease of the critical velocities v^{ID}_0, v^{FD}_0 and v^{SF}_0 with increasing half apex angle θ observed in Fig. 15 is mainly an effect of the increase in the maximum surface pressure with this angle. For the cylindrical projectile ($\theta = 0°$), the magnitudes of these velocities and the experimentally determined transition velocity v^*_0 are related as $v^{ID}_0 = 990$ m/s $< v^{SF}_0 = 1230$ m/s $< v^*_0 = 1442$ m/s $< v^{FD}_0 = 1690$ m/s. For the conical projectile ($\theta = 5°$) and copper-covered target, they are related as $v^{ID}_0 = 960$ m/s $< v^{SF}_0 = 1030$ m/s $\approx v^*_0 = 1028$ m/s $< v^{FD}_0 = 1500$ m/s. Thus, for the conical projectile the transition velocity is much lower relative to the three critical velocities than that of the cylindrical projectile, and it is close to the critical velocity v^{SF}_0 associated with the formation of surface failure observed both in the experimental tests and in the simulations. This means that the difference in transition velocity between the cylindrical and the conical projectile is too large to be explained by the increased surface pressure alone.

The main reason for the different modes of penetration of cylindrical and conical projectiles, and the large difference in transition velocity, is believed to be the radial growth of the surface load from the conical projectile. This growth has the effect that cone-shaped surface cracks are exposed to the radial flow of projectile material under rising surface pressure. Therefore, when the pressure is sufficiently high, projectile material penetrates into the cracks as seen in the experiments.

CONCLUSIONS

The aim of the research at FOI was to estimate the transition velocity for various loading conditions and combinations of projectiles and ceramic targets and to identify critical velocities which may serve as upper and lower bounds for the transition velocity. These critical velocities have been related to the material properties of the projectile and the target.

The experiments show that there is a distinct transition from interface defeat to penetration. Possibly, there exists a unique transition velocity for each combination of projectile, target material and target configuration. Tests with different projectile materials (WHA and Mo) indicate that the transition occurs at a critical surface pressure.

The ceramic material with the highest hardness shows the highest transition velocity, 2110 m/s (Syndie), but for one type of ceramic material (four grades of silicon carbide), the ranking in terms of hardness is not the same as that in terms of transition velocity. This indicates that hardness alone, which reflects the shear yield strength, is not sufficient for estimation of the relative performances of ceramic materials in terms of transition velocities. It follows that the transition velocity cannot be controlled by plastic flow alone. Figure 11(b) shows that an increase in transition velocity corresponds to an increase in fracture toughness, although the uncertainty in fracture toughness measurement is large. This observation indicates that, for this target configuration, fracture may have more influence than plastic flow on the transition from interface defeat to penetration. As a consequence, the observed transition velocities may not be the maximum ones achievable. By suppression of the initiation and propagation of cracks through increase of the confining pressure, it may be possible to increase the transition velocities for all materials tested.

The main reason for the different modes of penetration of cylindrical and conical projectiles, and the large difference in transition velocity, is believed to be the radial growth of the surface load from the conical projectile. This growth has the effect that the surface crack is exposed to the radial flow of projectile material under rising surface pressure. Therefore, when the pressure is sufficiently high, projectile material penetrates into the crack as seen in the experimental tests.

FUTURE RESEARCH

Even if the dwell phenomenon is well understood, some fundamental questions still needs further examination. In several of the studies at FOI, a complicated interaction between the flowing projectile material and macro-cracks has been observed at velocities close to the transition velocity. Infiltration of macro cracks by projectile material may enhance crack propagation substantially and contribute to the degradation of the ceramic target and thereby inhibit further dwell. This phenomenon is linked to the strong influence of the confining pressure on the transition velocity since the propagation of macro-cracks is sensitive to the confining pressure.

ACKNOWLEDGMENTS

This work was carried out at the Swedish Defence Research Agency FOI (former FOA), Weapons and Protection Division, as part of the research on long-rod projectiles and armour. The work was funded by the Swedish Armed Forces ([8] was also supported by the Army Research Laboratory (USA) through US Naval Regional Contracting Centre).

I would like to express my gratitude to my colleagues and especially to René Renström and Lars Westerling for their enthusiasm and will to share their knowledge at any time.

REFERENCES
1. G. E. Hauver, P. H. Netherwood, R. F. Benck and L. J. Kecskes. Ballistic performance of ceramic targets. *U. S. Army Symposium On Solid Mechanics*, USA (1993).
2. G. E. Hauver, P. H. Netherwood, R. F. Benck and L. J. Kecskes. Enhanced ballistic performance of ceramic targets. *19th. Army Science Conference*, USA (1994).
3. E. J. Rapacki, G. E. Hauver, P. H. Netherwood and R. F. Benck. Ceramics for armours- a material system perspective. *7th. Annual TARDEC Ground Vehicle Survivability Symposium*, USA (1996).
4. P Lundberg, L. Holmberg and B. Janzon, An experimental study of long rod penetration into boron carbide at ordnance and hyper velocities. *Proc 17th Int Symp on Ballistics*, South Africa: Vol 3, pp. 251-265 (1998).
5. L. Westerling, P. Lundberg and B. Lundberg, Tungsten long rod penetration into confined cylinders of boron carbide at and above ordnance velocities. *Int. J. Impact Engng*, 25, 703-714 (2001).
6. V. Wiesner, T. Wolf, P. Lundberg and Lars Holmberg. A study of the penetration resistance of titanium diboride using a new technique for time resolved terminal ballistic registrations. *Proc. 18th Int. Symp. on Ballistics*, USA, (1999).
7. P. Lundberg, R. Renström and B. Lundberg, Impact of metallic projectiles on ceramic targets: transition between interface defeat and penetration. *Int. J. Impact Engng*, 24, 259-275 (2000).
8. P. Lundberg, B. Lundberg, Transition between interface defeat and penetration for tungsten projectiles and four silicon carbide materials. *Int. J. Impact Engng*, 31, 781-792 (2005).
9. P. Lundberg, R. Renström, L. Holmberg. An experimental investigation of interface defeat at extended interaction times. *Proc. 19th. Int. Symp. Ballistics*, Vol. 3, 1463-1469 (2001).

10. P. Lundberg, R. Renström, B. Lundberg, Impact of conical tungsten projectiles on flat silicon carbide targets: transition from interface defeat to penetration. *Int. J. Impact Engng.* 32, 1842-1856 (2006).
11. O. Andersson, P. Lundberg, Rene Renström, Influence of confinement on the transition velocity of silicon carbide. *Proc. 23rd. Int. Symp. Ballistics*, Tarragona, Spain (2007) (to be published).
12. P. Lundberg, R. Renström, L. Westerling. Transition between interface defeat and penetration for a given combination of projectile- and ceramic material. In *Ceramic Armour by Design*, Edited by J.W. McCauley *et. al.*, Ceramic Transactions, Vol. 134 (2002).
13. R. Renström, P. Lundberg, B. Lundberg. Stationary contact between a cylindrical metallic projectile and a flat target under conditions of dwell. *Int. J. Impact Engng*, 30, 1265-1282 (2004).
14. T. J. Holmquist and G. R. Johnson. Response of silicon carbide to high velocity impact. *J. Appl. Phys.*, 91(9), 5858-5866 (2002).
15. G. R. Johnson and T. J. Holmquist. Response of boron carbide subjected to large strains, high strain rates and high pressure. *J. Appl. Phys.*, 85(12), 8060-8073 (1999).
16. G. R. Johnson and W. H. Cook. A constitutive model and data for metals subjected to large strains, high strain rates, and high temperatures. *Proc. 7th. Int. Symp. Ballistics*, 541-547 (1983).

THE INFLUENCE OF TILE SIZE ON THE BALLISTIC PERFORMANCE OF A CERAMIC-FACED POLYMER

Paul J Hazell
Cranfield University
Defence College of Management and Technology
Shrivenham, Oxfordshire, SN6 8LA, UK

Mauricio Moutinho
Brazilian Army
Rio de Janeiro, Brazil

Colin J Roberson and Nicholas Moore
Advanced Defence Materials Ltd.,
Rugby, Warwickshire, CV21 3XH, UK

ABSTRACT
    Silicon carbide square tiles of four sizes and two different thicknesses have been bonded to polycarbonate layers to evaluate their performance using a conventional DoP technique. Four tile sizes were tested: 85mm, 60mm, 50mm and 33mm. In each case the residual depth-of-penetration into the polycarbonate was recorded. It was found that if the core was shattered, the depth-of-penetration reduced considerably with increasing tile size. If the core remained intact after completely penetrating the ceramic then tile size had no effect on residual penetration. A simple elastic wave propagation model can be used to predict the experimental results quite well.

INTRODUCTION

    For multi-hit protection it is necessary to retain as much ceramic material intact as possible after each subsequent hit. One of the ways that this can be achieved is by reducing the tile-size such that that if one tile has been destroyed protecting against a single projectile, the exposed area to subsequent strikes is minimized. Bless and Jurick have conducted a probability-based analysis of such mosaics to determine how multi-hit probability varies with tile size [1]. De Rosset [2] has also studied such patterned armours to examine the probability of defeating automatic weapons fire. For these types of analysis there is a requirement to know how the ballistic performance is affected by the proximity of the impact to the tile edge.

    The size of the tile is also important for ballistic testing of the ceramic. Good reviews of the various techniques are provided by James [3] and Normandia and Gooch [4]. There are clear advantages in using small tiles, not only in the cost of the ceramic but also the cost of the backing materials. The question is: what is the minimum size of tile that can be used for ballistic testing so that the intrinsic ballistic properties of the material can be tested?

    This paper intends to explore these issues and, for a sintered silicon carbide, answer this question.

EXPERIMENTAL SETUP

    The Depth of Penetration (DoP) technique as described by Rozenberg and Yeshurun [5] was used to measure the ballistic performance of the ceramic tiles (see Figure 1). In this work, polycarbonate was chosen as the backing material instead of more commonly used materials such as RHA or aluminium. The use of polycarbonate, which is less resistant to ballistic penetration, has the advantage that any small differences in the ballistic performance of the tile will result in relatively large differences in DoP. Polymethylmethacrylate can also be used for these types of tests [6]. Polycarbonate is clear so that analysis of DoP can be done instantly without the requirement of X-Ray. It also has a similar impedance to the fibre composite used in light armour systems which leads to a more realistic trial than using a semi-infinite steel or aluminium backing. In these trials multiple polycarbonate tiles were used; each 100mm × 100mm ×12mm clamped together to form a semi-infinite target.

Two silicon carbides of varying sizes were tested: a direct sintered silicon carbide (Morgan AM&T PS-5000) and a commercially available liquid phase sintered (LPS) silicon carbide. Their measured properties are presented in Table I along with the properties of polycarbonate taken from [7]. The elastic properties of the silicon carbides were very similar. The densities were measured using a gas pycnometer and the longitudinal wave velocities, Young's modulus values and Poisson's ratios were measured ultrasonically using Panametrics' 5MHz longitudinal and shear-wave probes. The true hardness values (HV_0) were calculated from a series of micro-hardness tests at different loads using an Indentec HWDM7. The elastic impedance was calculated from the well-known relationship

$$Z = \sqrt{E\rho} \,. \tag{1}$$

The PS-5000 was tested in tiles of 6.35mm and 7.50mm thickness; the LPS silicon carbide tiles were 7.50mm thick. The tiles were cut to 33×33mm, 51×51mm, 66×66mm and 85×85mm. Five tiles for each ceramic type and geometry were tested at each different size, resulting in a total of twenty targets of each ceramic. Each ceramic tile was glued using Araldite AV4076-1 and HY4076 hardener mixed in the proportion of ten to four in weight. The surface of the polycarbonate was abraded in order to improve the gluing quality. A film of adhesive was applied on the ceramic surface which was then manually pressed against the polycarbonate and twisted until a continuous adhesive layer free of air bubbles was obtained. All the targets were glued and let to set at room temperature for at least 72 hours in an environment protected from light and moisture.

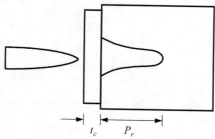

Figure 1: The DoP technique; t_c = tile thickness, P_r = depth-of-penetration.

Table 1: Measured properties of the silicon carbides used in this trial; the data for the polycarbonate is taken from [7].

Ceramic	ρ (kg/m^3)	c_L (m/s)	E (GPa)	v	Z (kg/m^{2}s)	HV$_0$
sSiC	3147	12021	427.0	0.16	36.0×10^6	2400
LPS SiC	3252	12111	446.0	0.17	38.1×10^6	2089
Polycarbonate	1190	2130	2.6	0.40	2.5×10^6	-

Figure 2 shows the 7.62mm AP "Sniper 9" round that was used for the ballistic tests. This projectile consists of a WC-Co cermet core placed in an Al cup and encased in a Cu-Zn jacket. The projectile's mass = 9.176g and measures 22.7mm in length and 7.8mm in diameter. The WC-Co core's mass = 5.556g and measures 22.3mm in length and 5.2mm in diameter with a 55° nose angle. The measured core hardness was 1292 ± 24 [HV2]. The average muzzle velocity of this bullet was 840m/s.

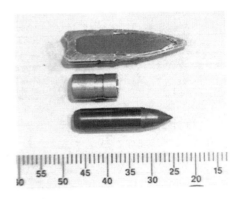

Figure 2: The 7.62mm AP Sniper 9 ammunition showing the WC-Co core, the aluminium cup and a sectioned bullet

After the tests, the polycarbonate was cut and the residual DoP of the projectile in the backing material was measured and recorded. The distance from the impact point to the borders of the ceramic tile were also measured and recorded.

RESULTS AND DISCUSSION
All shots were aimed at the centre of the ceramic tile but the actual impact position relative to the closest border was measured and recorded after each test. Figures 3 - 5 show the DoP-test results.

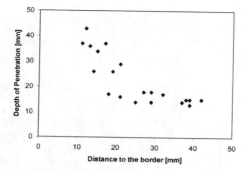

Figure 3: PS-5000 (7.5mm thickness) – DoP results.

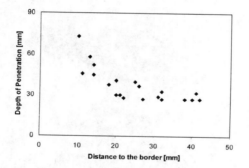

Figure 4: PS-5000 (6.35mm thickness) – DoP results.

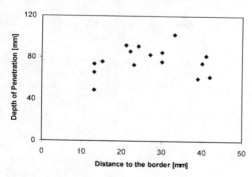

Figure 5: Liquid Phase Sintered SiC (7.5mm thickness) – DoP results.

From the DoP results presented in Figures 3 and 4 it can be seen that the PS-5000 ballistic performance was dependent of the distance from the impact point to the tile border. The performance increases as the impact occurs further from the border. The DoP for penetration at 10mm from the border was 40mm. This was almost three-times the DoP when the penetration was far from the border. For example, there was 15mm DoP when the distance was 45mm from the closest border. Beyond a critical distance (between 30 and 35mm) the DoP reaches a consistent value (allowing for experimental scatter). At this location, the material's intrinsic ballistic performance was measured due to the absence of border effects. It can also be noted that the border-independent DoP is about 15mm for the 7.5mm thickness tiles and 30mm for the 6.35mm thickness. This represents a gain of 100% in the ballistic performance for an extra 20% added thickness (and weight). On the other hand, the DoP results for the LPS silicon carbide seemed to be independent from the proximity of the border. The average DoP was 80 mm – performing much worse than the PS-5000 tiles of the same thickness. It has been noted by others [8] that the ballistic efficiency of liquid-phase-sintered SiC against WC-Co projectiles is significantly less than for sintered SiC (despite the similar impedance values) and therefore a thicker sample is required to shatter the projectile core.

Figure 6 shows a plot of tile size versus DoP for all three systems tested which summarises the test results. Each DoP data point for each tile size is the average measurements from five shots for the PS-5000 tiles and four shots for the LPS SiC.

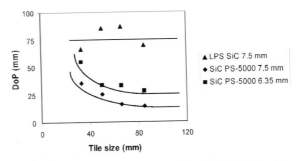

Figure 6: Summary of the ballistic test results plotted with tile size.

In order to investigate why the border proximity had a great effect on the SiC PS-5000 but no noticeable effect on the liquid phase sintered SiC, the fragments of the projectiles were collected and analysed. Figure 7 shows the projectile core after defeating the LPS SiC 7.5mm thickness tiles. It can be seen that no matter if the impact occurs close or far from the tile border, this system is not able to shatter the projectile core. A different behaviour is observed in the SiC PS-5000 7.5-mm thick tiles, as shown in Figure 8. In this target set-up the projectile core remains relatively intact if the shot hits the ceramic tile near to a border, but it is completely shattered when the impact occurs far from the borders of the tile. In the 85×85mm tiles only very small fragments of the core were found in the impact crater.

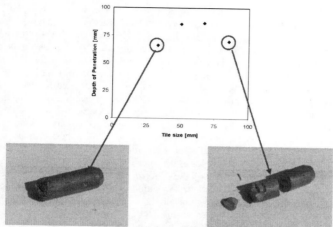

Figure 7: Projectile core after defeating the LPS SiC ceramic tiles

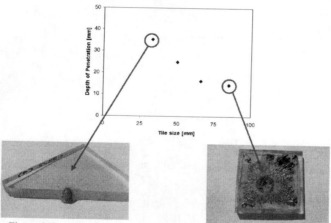

Figure 8: Projectile core after defeating the SiC PS-5000 ceramic tiles

The shattering of the core during impact and penetration plays an important role in the ballistic performance of the material. A projectile that is reduced to smaller fragments will be much less effective in penetrating the target and defeating the armour system than an intact projectile. For the PS-5000, the core penetrates relatively intact when the impact point is close to the border. When the point of impact is far from the border, the core is shattered (see Figure 8). This explains the drop-off in DoP with this ceramic. The same does not occur with the liquid phase sintered tiles because there is little difference in the projectile's morphology during penetration regardless of the proximity of impact to a border. In this case, the projectile penetrates in a rigid-body manner.

The PS-5000 behaviour can be explained by the stress-wave propagation in the ceramic that expands in a hemi-spherical fashion until it interacts with boundaries. During the ballistic impact a compressive stress wave is delivered to both the ceramic and the projectile. This wave consists of a fast-moving elastic portion and a slower-moving inelastic portion that disperses with distance. The first interface that is encountered by the wave is at the ceramic / polycarbonate interface. Due to the relatively low impedance of the adhesive layer between the ceramic and the polycarbonate, much of the wave is reflected as a tensile wave that releases the state of compression and causes damage. However, at this point in time the intact material surrounding the damaged zone provides inertial confinement. Similarly the radially expanding portion of the wave eventually encounters a border and due to the presence of a free surface, is completely reflected as a tensile wave. Further damage to the ceramic then ensues. It is this portion of the wave that is responsible for the drop-off in ballistic performance when the impact site occurs closer to the border. Making the assumption that the wave is moving in a bounded medium such as a 7.5-mm diameter bar instead of a 7.5-mm thick tile, the speed of the elastic wave can be calculated from the relationship:

$$c_0 = \sqrt{\frac{E}{\rho}} \tag{2}$$

and is 11.64km/s for the PS-5000 and 11.71km/s for the liquid phase sintered SiC.

Cline and Wilkins [9] identified the importance of the impedance of the ceramic for defeating a projectile. Recent work [10] has suggested that there is a critical thickness required to shatter a WC-Co projectile and this correlates with the elastic impedance of the tile. For this type of sintered silicon carbide it has been found that this critical thickness is ~4.3mm for a 51mm×51mm square tile. For a larger tile, it is likely to be slightly thinner.

The ability of the ceramic to shatter the core is related not only to the magnitude of the shock stress (which, in turn, is related to the relative impedances of the target and projectile) but also to the duration of the shock stress. This latter property is determined by the length of time the ceramic remains intact below the penetrating projectile. The compressive stress wave formed in the ceramic travels inside the material until reaching a free surface where it is reflected back as a much more damaging tensile stress wave. If the distance to the border is small the tensile wave may reach the impact area while the projectile has not yet been shattered. The already broken ceramic material ahead of the projectile is released by the arrival of the tensile wave and it is not able to sustain the stress for the duration necessary to induce shattering. As a result, the projectile penetrates in a rigid-body manner.

The time taken to penetrate 4.3mm of silicon carbide assuming an impact velocity of 840m/s is approximately 5μs. This assumes that there is no deceleration of the projectile. This time represents the time required for the projectile to maintain contact with the ceramic and is the limiting case. For the 6.35mm and 7.5mm the time will be longer. Using Equation 2 the elastic wave would have travelled approximately 60mm in this time. This represents an impact of 30mm from the border and allowing for the simplification of using this model, is consistent with the result that the drop-off in ballistic performance plateaus between 30mm and 35mm.

CONCLUSIONS

The effect of tile size has been studied on two silicon-carbide-faced polymer targets. It was found that:

- The measured DoP after completely penetrating 6.35-mm and 7.5-mm thick Morgan AM & T PS-5000 is dependent on the tile size.
- For both thicknesses of this tile the effect of the border was insignificant at a proximity of impact of approximately 30-35mm. Consequently, the minimum square tile-size that should be used so that the intrinsic ballistic properties of this material can be tested is 70mm × 70mm.
- The PS-5000 out-performed the LPS SiC ballistically in the DoP-test configuration.
- The LPS SiC 7.5-mm thick tile showed no variation of ballistic performance with tile size. The projectile penetrated in a rigid-body fashion. Had a thick enough tile been used to shatter the projectile core it is expected that the size-effect behaviour would become apparent in the results as it did with the PS-5000.
- Border effects are important when the ceramic material is able to induce shattering in the projectile core. In this study, border effects are not important when the armour is overmatched and the projectile penetrates in a rigid-body fashion.
- The border-effect can be predicted by a simplified elastic wave-speed model.

ACKNOWLEDGEMENTS

This work was carried out during Capt Moutinho's study for a Forensic Engineering and Science MSc at Cranfield University. We would like to thank the Brazilian Army for funding his studies. We would also like to acknowledge Mr David Miller and Mr Adrian Mustey for their technical assistance. Finally, our thanks to Morgan AM&T for supplying the samples and funding other materials used in the project.

REFERENCES

[1] S. J. Bless and D. L. Jurick, "Design for multi-hit protection", *Int. J. Impact Eng.*, **21** (10), 905-8 (1998).

[2] W. S. de Rosset, "Patterned armor performance evaluation", *Int. J. Impact Eng.*, **31**, 1223-34 (2005).

[3] B. James, "Depth of penetration testing". In *Ceramic Armor Materials by Design, Ceramic Transactions*, Vol.134, J. W. McCauley *et al.* (Eds), 165-72 (2002).

[4] M.J. Normandia and W.A. Gooch, "An overview of ballistic testing methods of ceramic materials". In *Ceramic Armor Materials by Design, Ceramic Transactions*, Vol.134, J. W. McCauley *et al.* (Eds), 113-38 (2002).

[5] Z. Rozenberg and Y. Yeshurun, "The relationship between ballistic efficiency and compressive strength of ceramic tiles", *Int. J. Impact Eng.*, **7** [3], 357-62, (1988).

[6] I. D. Elgy, I. M. Pickup and P. L. Gotts, "Determination of the critical components of a complex personal armour rigid plate". *In the proceedings of Personal Armour Systems Symposium (PASS2006)*. Royal Armouries Museum, Leeds, UK, 18th -22nd September (2006).

[7] J. C. F. Millett and N. K. Bourne, "Shock and release of polycarbonate under one-dimensional strain". *J. Mater. Sci.*, **41**, 1683-90, (2006).

[8] D. Ray, R. Flinders, A. Anderson and R. Cutler, "Effect of room-temperature hardness and toughness on the ballistic performance of SiC-based ceramics," *Ceram. Sci. Eng. Proc.,* **26** [7] 131-42 (2005).

[9] C. F. Cline and M. L. Wilkins, "The importance of material properties in ceramic armor", *Proceedings of the Ceramic Armour Technology Symposium,* USA, pp. 13-18, January (1969).

[10] P.J. Hazell and C.J. Roberson, "On the critical thickness of ceramic to shatter WC-Co bullet cores". *In the proceedings of the 22nd International Symposium on Ballistics,* Vancouver, Canada, 14-18 November 2005.

STATIC AND DYNAMIC INDENTATION RESPONSE OF FINE GRAINED BORON CARBIDE

Ghatu Subhash,[1*+] Dipankar Ghosh[2] and Spandan Maiti[3]

[1]Department of Mechanical and Aerospace Engineering
University of Florida, Gainesville, FL 32611, USA
[2]Department of Materials Science and Engineering
[3]Department of Mechanical Engineering-Engineering Mechanics
Michigan Technological University, Houghton, MI 49931-1295, USA

ABSTRACT

Fine grained boron carbide (B_4C) ceramic disks of different grain sizes were consolidated using plasma pressure compaction ($P^2C^®$) method. Static and dynamic indentations were conducted to determine the strain rate sensitivity of indentation hardness and fracture toughness as a function of grain size. It was noted that the indentation-induced cracks during static indentation were transgranular in nature with zig-zag pattern within each grain. This zig-zag pattern is speculated to follow the cleavage planes in the ceramic and results in the formation of intermittent parallel crack facets. On the other hand, indentation-induced cracks under dynamic indentations showed relatively straight cracks with no zig-zag pattern within the grains. A finite element based computational framework based on fracture mechanics of brittle materials has been developed to model the observed zig-zag crack propagation pattern under static indentations. Cohesive zone modeling in conjunction with a novel formulation termed as generalized cohesive element method has been employed to model and resolve the intragranular crack motion in these ceramics. Preliminary results of simulations are presented and future research directions are summarized.

1. INTRODUCTION

Boron carbide (B_4C) is one of the lowest density (2.52 g/cm^3) structural ceramics with high melting point (2450°C), high elastic modulus (450 GPa) and high hardness (HV 25-35 GPa,) next only to diamond and cubic boron nitride. It finds numerous applications[1] at room and high temperatures for wear resistant ceramic components and also as potential body armor. Boron carbide can be sintered by conventional techniques such as pressureless sintering[1-6], hot-pressing[1,7] and hot isostatic pressing (HIP)[1,8]. All these processing techniques require high sintering temperature (>2000°C) and long consolidation time (on the order of hours) to produce dense boron carbide. The above conditions invariably result in coarse grained microstructure. Besides, use of sintering aids can reduce fracture strength moderately[6].

*Corresponding author: subhash@mtu.edu, Ph: (906) 487-3161, Fax: (906) 487-2822
+Member- The American Ceramic Society
Supported by the National Science Foundation, USA

In the current work, boron carbide has been consolidated using a novel non-conventional method known as plasma pressure compaction ($P^2C^®$) technique. Earlier studies[9] employing P^2C technique to produce boron carbide utilized sintering additives such as graphite, alumina and titanium diboride. The processing temperature was reduced to 1650°C and a consolidation time of 5 min was used to obtain a B_4C disk of 97% of theoretical density. In the current study, we have sintered boron carbide without use of any sintering additives to produce theoretically dense and fine-grained compacts with various grain sizes.

Due to its high hardness and low density, B_4C has been a potential a candidate for armor ceramic[1,12]. Flexural strength, fracture toughness,[1-5] and nanoindentation response[10,11] were investigated in the literature. To evaluate its effectiveness under dynamic loads, high strain rate experiments using split Hopkinson pressure bar, plate impact and projectile impact experiments were conducted. But these experiments are relatively expensive, time consuming and require large size specimens.[12-15] Therefore, in the current investigation dynamic indentation experiments[16] were conducted to evaluate the mechanical behavior of boron carbide under high strain rate conditions. To compare the strain rate sensitivity of hardness, static indentation experiments were also performed. Microstructural analysis was later performed on the specimens to identify the influence of strain rate on indentation induced crack patterns. A 2-D plane strain numerical framework based a novel generalized cohesive element method in conjunction with the conventional cohesive volumetric finite element (CVFE) model[17,18] has been developed to simulate the observed fracture behavior and gain further insight.

2. EXPERIMENTAL

Boron carbide powder with 800 nm particles was used as the starting material, see Fig. 1. Consolidation of the powder was performed under high vacuum (200-300 mTorr) without the use of any sintering aids in the $P^2C^®$ method[9,19]. The powder was subjected to low voltage (5 V) high current density (4400 amp/cm^2) electric field to cause "Joule" heating along the inter particle contact areas. Sintering is facilitated by application of external pressure (88 MPa) at a temperature of 1750°C. Sintered disks of 51 mm diameter and 6.4 mm thickness, as shown in Fig. 2, were produced. By varying the consolidation times for 2 min, 5 min and 30 min, respectively, B_4C disks with different grain sizes were produced. Grain size measurements were performed using optical micrographs of the polished and the etched surfaces. Line intercept method (ASTM standard E 112-96) was used to estimate the average grain size. Etching was performed electrolytically using 1% KOH solution at a current density of 0.03 amp/cm^2 for 30 seconds. Density measurements using Archimedes method revealed that only 96% (2.42 g/cc) theoretical density was achieved in the disk consolidated for 2 min. By increasing the consolidation time to 5 min nearly full theoretically dense (2.5 g/cc, ~ 99.2%) disk were obtained. Further increase in time to 30 min caused only grain coarsening.

Fig.1. Transmission electron micrograph of the starting boron carbide powder particles.

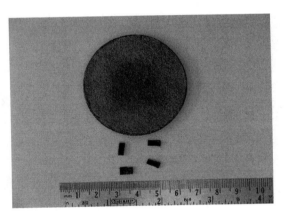

Fig. 2. Consolidated boron carbide disk and test samples for indentation.

Specimens of rectangular cross section of 4mm × 3mm and of length 25mm were used to investigate the rate dependent indentation fracture characteristics on the loading surface. The 4mm × 3mm loading surfaces of each specimen were polished metallographically to perform static and dynamic Vickers indentation experiments. Static indentations were performed at three different loads of 2.94 N (300 gm), 4.9 N (500 gm) and 9.8 N (1000 gm) for 15 seconds. At each load, around 15 indentations were performed. A dynamic indentation tester[16], shown in Fig. 3, was employed for dynamic hardness measurements at comparable load levels. In this technique, utilizing the principal of elastic stress wave propagation in a slender rod, the desired load can be delivered within 100 - 150 μs duration. This technique parallels the static indentation technique as the indentation load can be measured directly without any prior knowledge of material properties. The dynamic hardness tester consists of a slender rod with a Vickers indenter mounted at one end and a momentum trap (MT)[16,20] assembly at the other end. For indentation load measurements, a high frequency load cell is mounted on a rigid base as shown in Fig. 3. The specimen is sandwiched between the diamond indenter and the load cell. A striker bar is launched from a gas gun (not shown in the figure) to generate an elastic stress pulse in the incident bar (at the MT end). Magnitude of the stress pulse depends on the striker bar velocity and its duration depends on the length of the striker bar. The stress pulse then travels through the incident bar and results in an indentation on the specimen. The MT assembly ensures that a single compressive pulse reaches the indenter and thus only a single indentation is performed on the specimen. The dynamic hardness is calculated based on indentation diagonal size and load. For ceramic specimens, the indentation induced strain rate was estimated to be around 1000/s[16]. An average of 25-30 tests per specimen were conducted for each grain size at load ranges similar to those under static indentation.

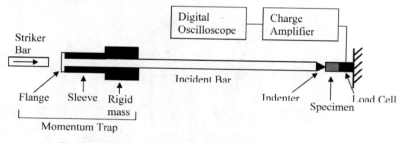

Fig. 3. Schematic of experimental setup for dynamic indentation hardness measurements.

Indentation induced cracks from the corners of the static and dynamic indents were measured for fracture toughness (K_{IC}) calculations. When the data between the half diagonal crack length c and load P were plotted to fit the equation of type $c = AP^n$, the n values were in the range of 0.62-0.69. Thus it was determined that these cracks were half-penny cracks[21] and therefore, Evans and Charles[22] fracture toughness equation, $K_{IC} = 0.0824P/c^{1.5}$ was used for static and dynamic fracture toughness measurements. To study the influence of strain rate on the indentation induced crack patterns, static and dynamic indentation experiments were also conducted on the polished and etched boron carbide specimens. Then scanning electron microscopy was performed to observe the crack patterns.

3. RESULTS

3.1 Density and microstructure

Boron carbide disks sintered at 1750°C for 2 min achieved only 96% of the theoretical density. The porosity associated with this disk is clearly seen on the fracture surfaces shown in Fig. 4(a). Upon increasing the sintering time almost theoretically dense sintered boron carbide disks were obtained, as shown in Fig. 4(b) and 4(c). Optical micrographs of polished and etched specimen of the above disks are shown in Fig. 5. The average grain size measurements from line intercept method revealed that the grain sizes are 1.6 μm, 2 μm and 2.7 μm, respectively for disks consolidated for 2 min, 5 and 30 min. Darker regions in the micrograph represent mostly grain pull-outs during polishing.

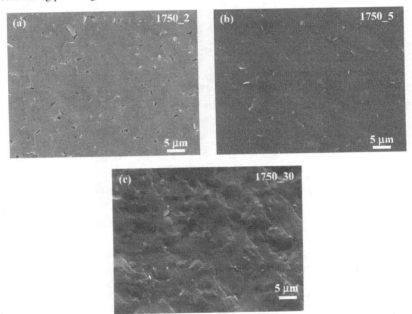

Fig. 4. SEM micrographs of sintered boron carbide samples at 1750°C for various consolidation times: (a) 2 min; (b) 5 min and (c) 30 min.

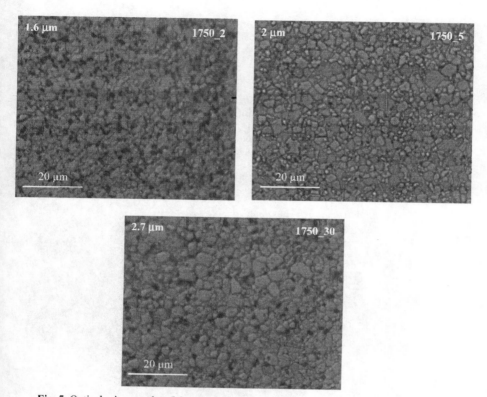

Fig. 5. Optical micrographs of the sintered (at 1750 °C) and etched boron carbide samples. The average grain sizes are indicated on the optical micrographs.

3.2 Comparison of static and dynamic indentation hardness

The results of static and dynamic hardness measurements for the three grain sizes are presented in Fig. 6. Both static and dynamic hardness values showed a wide scatter for all the grain sizes. In the tested load range, the 2 μm and the 2.7 μm grain size boron carbide disks showed similar average static hardness values (27.45 ± 2 GPa and 27.45 ± 2.5 GPa) where as the 1.6 μm grain size boron carbide showed an average static hardness value of 25.41 ± 1.0 GPa. In general, compared to static hardness, a lower dynamic hardness was observed for all the three grain sizes. The 2 μm grain size boron carbide showed the highest average dynamic hardness (26.15±2.5 GPa) followed by the 2.7 μm (23.37±3.0 GPa) and the 1.6 μm (18.17±2.0 GPa) grain size specimens.

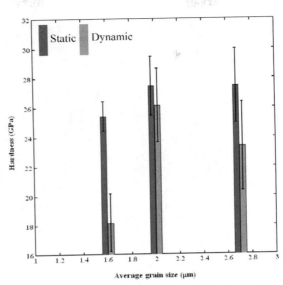

Fig. 6. Comparison of static and dynamic hardness values
for three grain sizes of boron carbide.

3.3 Indentation crack lengths and fracture toughness

Optical micrographs in Fig. 7 show a comparison of the damage on the top surface of the indented regions of 2.7 μm grain size specimen at an indentation load of 2.94 N. In general the damage is more severe under dynamic indentation compared to static indentation. Crack length measurements revealed a linear increase in half-median crack length (c) with indentation load. The average crack lengths were slightly longer under dynamic loads compared to static loads for all the grain sizes as can be observed from Fig. 8. Accordingly, a lower fracture toughness (utilizing Evans and Charles equation[22] described in Section 2) was observed under dynamic loads compared to static loads (see Fig. 9).

Clearly, the hardness and the fracture toughness results showed a loss of strength under dynamic loading compared to static loading. These results support other experimental results at high strain rates where similar drop in mechanical properties was observed.[13-15] To investigate the underlying cause for loss of hardness and fracture toughness, further studies using Raman spectroscopy were conducted.[23] It was observed that dynamic indentation caused a greater level of structural changes (localized amorphization) in boron carbide compared to static indentation and thus a lowering in strength at high strain rate was resulted.

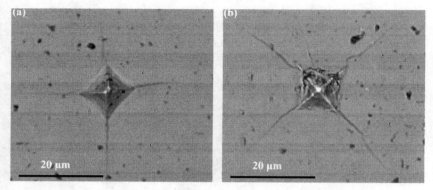

Fig. 7. Optical micrographs of (a) static and (b) dynamic indents for 2.7 μm grain size boron carbide at 2.94 N. Note longer cracks and more severe damage in dynamic indent.

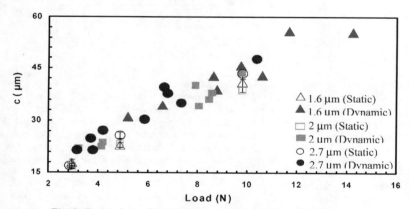

Fig. 8. Comparison of half-diagonal crack length (c) with indentation load.

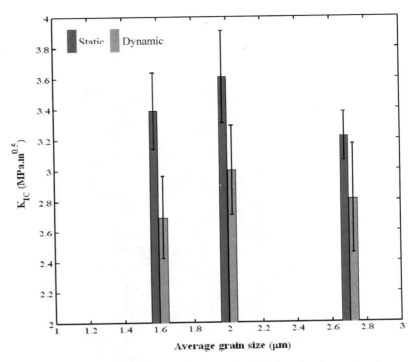

Fig. 9. Comparison of static and dynamic indentation fracture toughness for three different grain sizes of boron carbide.

3.4 Indentation induced crack morphology

SEM micrographs, shown in Fig. 10, reveal the microscopic crack pattern under static and dynamic indentations. Although the fracture mode was primarily transgranular under both type of loading conditions, the static indentation-induced crack exhibited intermittent zig-zag type crack path where as the dynamic indentation-induced crack revealed a straight crack path. To rationalize this behavior, we have initiated a computational frame work as will be described in the Section 4.

The observed intermittent zig-zag type crack pattern within the grains under static indentations has been attributed to the formation of steps or cleavage facets on the fracture surface. This type of unusual transgranular fracture has been speculated as the competition between crack growth direction and the energetically favorable cleavage plane. It has been proposed that if the crack propagates on a plane which is not favorably oriented with respect to loading direction, then the crack can deflect to some other plane which is energetically more favorable. Under these circumstances, traces of the crack path will not be smooth like typical transgranular fracture. Rather several steps can be observed on the fracture surface. Schneibel et al[24] also noted similar cleavage steps on the fracture surface (011) of Al3Sc which supports the proposed mechanism by the authors of this paper. On the other hand, the observed straight crack path during dynamic loading has been attributed to the typical transgranular fracture. The rapid input of energy and the material inertia at high strain rate loading inhibit local crack deflection and sometimes results in multiple cracks to emanate from the indentation corners. More in depth studies are under progress to rationalize the observed behavior in the crack patterns.

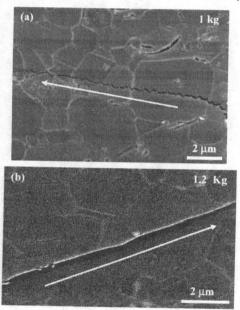

Fig. 10. Micrograph of indentation induced crack patterns in boron carbide under (a) static and (b) dynamic indentations. Indentation loads are indicated.

4. NUMERICAL FRAMEWORK

The authors have developed a 2-D plane strain numerical framework to simulate the static indentation induced crack patterns mentioned in the previous sections. Crack propagation on a single cleavage plane has been simulated using the cohesive volumetric finite element (CVFE) model.[17,18] while to simulate the off-cleavage plane fracture (crack deflection from the cleavage plane), a novel formulation termed as generalized cohesive element (GCE) method[25] has been introduced. In the following paragraphs the concept of generalized cohesive elements along with its finite element implementation will be discussed briefly.

4.1 Conventional cohesive model for fracture

In the standard cohesive element[17,18] technique for computational fracture simulation, conventional volumetric elements and cohesive elements are combined to capture the constitutive and failure response of materials, respectively. Standard finite elements with constant strain triangles were used to capture the constitutive response of the bulk material whereas 4-noded cohesive elements (as shown in Figure 11) are placed along the boundary of the volumetric elements to model the interelement fracture. For the current study, cohesive elements placed along the boundary of the volumetric elements are characterized by a bi-linear rate-independent but damage dependent traction-separation law that correlates the displacement jump vector (Δ) and cohesive traction vector ($\mathbf{T}$) acting across the cohesive surfaces, Γ_c. Normal and tangential cohesive tractions, denoted by T_n and T_t, respectively, can be calculated from normal and tangential crack opening displacements Δ_n and Δ_t across the cohesive surfaces using the following relationships:

$$T_n = \frac{s}{(1-s)} \frac{\Delta_n}{\Delta_{nc}} \frac{\sigma_{max}}{s_{init}}$$

(1)

$$T_t = \frac{s}{(1-s)} \frac{\Delta_t}{\Delta_{tc}} \frac{\tau_{max}}{s_{init}}$$

(2)

where normal and tangential components are indicated by the subscripts n and t. σ_{max} and τ_{max} correspond to maximum tensile and shear failure strengths, while Δ_{nc} and Δ_{tc} denote the critical displacement jumps in normal and tangential directions, respectively. The parameter s denotes a monotonically decreasing damage parameter which quantifies the evolution of damage in the material and couples mode I and mode II failure.[17,18] Cohesive elements start to fail as s gradually decreases from its initial value s_{init} (generally taken as 0.98) due to the separation across cohesive surfaces. Cohesive elements fail completely when s reaches zero.

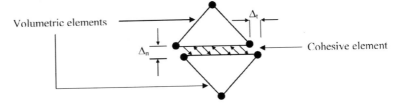

Volumetric elements

Cohesive element

Fig. 11: Normal and shear displacements of a cohesive element placed between two volumetric elements.

4.2 Generalized cohesive element technique

A major drawback of the cohesive technique is that the propagating crack is provided with a predetermined number and orientation of crack paths restricting resultant crack motion. Also, crack path becomes mesh dependent as they are provided only at the element edges (see Fig. 11). To represent arbitrary crack motion in its truest sense, we need to incorporate interelement as well as intra-element crack paths such that crack propagation becomes mesh independent. But standard finite element formulations in conjunction with conventional cohesive methodology cannot account for truly arbitrary crack paths as the displacement field becomes discontinuous within the element whenever intraelement crack motion is present. We have developed a generalized cohesive element technique which can take care of inter-element as well as intra-element crack motion. Basic idea is to relate displacements of the nodes in the vicinity of the crack to the displacement jump at the crack path. Topology of the finite element mesh determines the nodes in the vicinity of a crack.

Arbitrary crack propagation through the finite elements i.e., intraelement crack propagation can be modeled by two formulations: elements with embedded discontinuities and extended finite elements methods.[26,27] Recently, Mergheim et al[28] and Song et al[29] have modeled intraelement crack propagation using the concept of virtual[30] or phantom node algorithm.[29] In the current manuscript, we will refer this concept as virtual node algorithm. In this formulation, a deformed finite element containing the crack is replaced by two superimposed elements with additional virtual nodes while the original nodes and the material embedded within the original element are distributed within the two elements as shown in Fig. 12 (a). It can be seen that a part of the total material goes to each new element conserving total mass. Within each triangular element, the area surrounded by the solid lines contains the original material where as rest of the element is empty only containing the virtual nodes. If these two triangular elements are superimposed, the original deformed triangular element will be obtained. The displacement field of the original element is now considered to be composed of two independent displacement fields corresponding to each new element. Each displacement field is continuous within the respective element and the displacement jump along the discontinuity is measured as the difference of the two displacement fields.

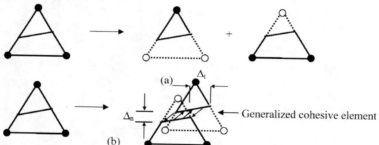

Fig. 12: (a) Superimposed elements and the separated elements. Areas enclosed by the dotted lines do not contain any material. Empty circles indicate the virtual nodes.
(b) generalized cohesive element between two superimposed volumetric elements.

Generalized cohesive elements are introduced whenever principal stress in an element exceeds the failure strength of the material. These elements follow the same failure law as already mentioned in the section on CVFE model enabling them to represent mixed mode fracture (see Fig. 12 (b)). Once the displacement jumps are calculated along the crack, cohesive tractions can be found from Eq. (1) and (2).

4.3 Numerical studies

To rationalize the experimentally observed static indentation induced crack patterns, some preliminary numerical studies have been conducted on a finite element mesh containing a single cleavage plane (consists of cohesive elements and with an orientation of 30°) whereas rest of the domain consists of CST elements (see Fig. 13). The computational domain is fixed at the right edge while displacement Δ has been applied to the left edge. In these preliminary studies, influence of failure strength of material on crack propagation through cleavage and off-cleavage plane has been investigated. The material properties applied for the volumetric elements are E = 460 GPa and ν = 0.17 whereas for the cohesive and generalized cohesive elements are σ_{max} = 250 MPa. τ_{max} = 250 MPa, Δ_{nc} = 50 µm and Δ_{tc} = 50 µm.

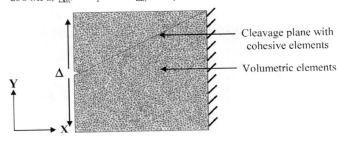

Fig. 13: Finite element mesh containing a single cleavage plane with an orientation of 30°.

4.3.1 Influence of failure strength of material

Figure 14 shows the contour plots of two case studies where only the failure strength of the material was varied whereas all other material properties including fracture toughness were kept the same. For low failure strength of the material, the crack deflected from the cleavage plane early towards energetically more favorable orientation, while an increase of the failure strength from 335 to 400 MPa restricts the crack path only along the cleavage plane. Since the cleavage plane is situated at an angle (30°) with respect to the loading direction (in the direction of maximum principal stress), crack propagation on cleavage plane was not favorable. Therefore, as the maximum elemental principal stress exceeded the failure strength e.g., case-I, off-cleavage plane cracking initiated and the crack propagated in the direction of applied load i.e., energetically more favorable direction (Fig. 14 (a)). But for higher failure strength (case-II), the crack still continued on the cleavage plane as the maximum elemental principal stress did not exceed the failure strength. The expanded view in Fig. 14 (c) reveals the intraelement crack path due to the introduction of generalized cohesive elements described before. Thus, the crack propagation was energetically more favorable for case-I compared to case-II and thus resulted in a larger final crack length for the former case compared to the later case.

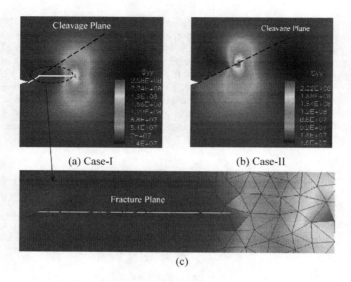

(a) Case-I   (b) Case-II

(c)

Fig. 14: Crack path simulations for failure strengths (a) 335 MPa and (b) 400 MPa. (c) Close up view of intraelement crack path of (a).

Our future work will focus on a detailed parametric study that will include the effect of fracture strength, different cleavage plane orientations, fracture toughness etc to fully understand the competition between the energetics of macroscopic crack growth direction and local cleavage plane orientation.

CONCLUSIONS

1. Employing plasma pressure compaction ($P^2C^®$) technique, near theoretically dense boron carbide ceramic disks were produced at significantly lower processing time and temperatures than in traditional sintering methods. Lowering in sintering temperature and time resulted in the development of fine-grained microstructure.

2. Dynamic indentations resulted in a consistent decrease of hardness and fracture toughness as well as greater extent of damage compared to static indentations for all the grain sizes. The significant decrease in dynamic hardness for 1.6 μm grain size boron carbide was attributed to the presence of higher level of residual porosity compared to other grain sizes.

3. Both type of indentations exhibited transgranular fracture pattern. But the crack path under static indentation also showed an intermittent zig-zag pattern within the grains which was not observed under dynamic indentation. The observed crack pattern under static loading has been correlated to the formation of cleavage steps on fracture surface.

4. A 2-D finite element framework has been developed to model the transgranular cracking observed under static indentations.

5. The preliminary numerical studies capture the competition between on- and off-cleavage plane crack propagation depending on the material properties and their relative orientations.

ACKNOWLEDGEMENT:
This work is funded by a grant from the US NSF (grant# CMS-0324461) with Dr. Ken Chong as the program manager.

REFERENCES
[1]F. Thevenot, "Boron Carbide-A Compressive Review," *J. Eur. Ceram. Soc.*, 6, 205-225 (1990).
[2]H. W. Kim, Y. H. Koh and H. E. Kim, "Densification and Mechanical Properties of B_4C with Al_2O_3 as a Sintering Aid," *J. Am. Ceram. Soc.*, 83 [11] 2863-2865 (2000).
[3]L. S. Sigl, "Processing and Mechanical Properties of Boron Carbide Sintered with TiC," *J. Eur. Ceram. Soc.*, 18, 1521-1529 (1998).
[4]Z. Zahariev and D. Radev, "Properties of Polycrystalline Boron Carbide Sintered in the Presence of W_2B_5 without Pressing," *J. Mat. Sci. Lett.*, 7 [7] 695-696 (1988).
[5]H. Lee and R. F. Speyer, "Hardness and Fracture Toughness of Pressureless-Sintered Boron Carbide (B_4C)," *J. Am. Ceram. Soc.*, 85 [5] 1291-1293 (2002).
[6]H. Lee and R. F. Speyer, "Pressureless Sintering of Boron Carbide," *J. Am. Ceram. Soc.*, 86 [9] 1468-1473 (2003).
[7]R. Angers and M. Beauvy, "Hot-Pressing of Boron Carbide," *Ceram. Int.*, 10 [2] 49-55 (1983)
[8]H. T. Larker, L. Hermansson and J. Adlerborn, "Hot isostatic pressing and its applicability to silicon carbide and boron carbide," *Mater. Sci. Monogr.*, 38A, 795-803 (1987).
[9]B. R. Klotz, K. C. Cho and R. J. Dowding, "Sintering Aids in the Consolidation of Boron Carbide (B_4C) by the Plasma Pressure Compaction (P^2C) Method," *Mater. Manufac. Process*, 19, 631–639 (2004).
[10]V. Domnich and Y. Gogotsi, "Nanoindentation and Raman Spectrosopy Studies of Boron carbide Single Crystals," *Appl. Phys. Lett.*, 81 [20] 3783-3785 (2002).
[11]D. Ge, V. Domnich, T. Juliano, E. A. Stach and Y. Gogotsi, "Structural Damage in Boron Carbide Under Contact Loading," *Acta Mater.*, 52, 3921-3927 (2004).
[12]C. Kaufmann, D. Cronin, M. worswick, G. Pageau and A. Beth, "Influence of Material Properties on the Ballistic Performance of Ceramics for Personal Body Armor," *Sock and Vibration* 10, 51-58 (2003).
[13]M. Chen, J. W. McCauley and K. J. Hemker, "Shock-Induced Localized Amorphization in Boron Carbide," *Science*, 299 7 March, 1563-1566 (2003).
[14]D. E. Grady, "Dynamic Properties of Ceramic Materials," Sandia National Laboratories Report, SAND 94-3266, Sandia National Laboratories, Albuquerque, NM, (1995).
[15]D. Dandekar, "Shock Response of Boron Carbide," ARL-TR-2456, Army Research Laboratory, Aberdeen Proving Ground, Aberdeen, MD, (2001).
[16]G. Subhash, "Dynamic Indentation Testing," *ASM Hand Book*, 8, 519 – 529 (2000).
[17]P. H. Geubelle and J. S. Baylor, "Impact-induced delamination of composites: a 2D simulation," *Composites Part B* 29B, 589–602 (1998).
[18]S. Maiti, P. H. Geubelle, "A cohesive model for fatigue failure of polymers," *Engineering Fracture Mechanics* 72 691–708 (2005).

[19]T. S. Srivatsan, R. Woods, M. Petraroli and T. S. Sudarshan, "An Investigation of the Influence of Powder Particle Size on Microstructure and Hardness of Bulk Samples of Tungsten Carbide," *Powder Tech.*, 122 [1] 54–60 (2002).

[20]S. Nemat-Nasser, "Recovery Hopkinson Bar Techniques," *ASM Hand Book*, 8, 477–487 (2000).

[21]D. K. Shetty, I. G. Wright, P. N. Mincer and A. H. Clauer, "Indentation Fracture of WC-Co Cermets," *J. Mater. Sci.*, 20 [5] 1873-1882 (1985).

[22]C. B. Ponton and R. D. Rawlings, "Vickers Indentation Fracture Toughness Test Part 1 Review of Literature and Formulation of standardized Indentation Toughness equations," *Mater. Sic. and Tech.*, 5 865-872 (1989).

[23]D. Ghosh, G. Subhash, T. S. Sudarshan, R. Radhakrishnan and X.-L. Gao, "Dynamic Indentation Response of Fine-Grained Boron Carbide," *J. Am. Ceram. Soc.*, 2007 (accepted).

[24]J. H. Schneibel and P. M. Hazzledine, "The crystallography of cleavage fracture in Al3Sc," *J. Mater. Res.*, 7 [4] 868 – 875 (1992).

[25]S. Maiti, D. Ghosh and G. Subhash, "A Generalized Cohesive Element Technique for Arbitrary Crack Motion," in preparation.

[26]I. Babuska and J. M. Melenk, "The partition of unity method," *IJNME*, 40 727-758 (1997).

[27]T. Belytschko, H. Chen, J. Xu and G. Zi, "Dynamic crack propagation based on loss of hyperbolicity with a new discontinuous enrichment," *IJNME*, 58 1873-1905 (2003).

[28]J. Mergheim, E. Khul and P. Steinmann, "A finite element method for the computational modeling of cohesive cracks," *Int. J. Numer. Meth. Engng.* 1-3 (2000).

[29]J. H. Song, P. M. A. Areias and T. Belytschko, "A method for dynamic crack and shear band propagation with phantom nodes," *Int. J. Numer. Meth. Engng.* 868 – 893 (2006).

[30]N. Molino, Z. Bao and R. Fedkiw, "A Virtual Node Algorithm for Changing Mesh Topology During Simulation," *International Conference on Computer Graphics and Interactive Techniques*, 385-392 (2004).

HIERARCHY OF KEY INFLUENCES ON THE BALLISTIC STRENGTH OF OPAQUE AND TRANSPARENT ARMOR

Andreas Krell
Fraunhofer Institut für Keramische Technologien und Systeme (IKTS)
Winterbergstraße 20
D - 01277 Dresden, Germany

Elmar Strassburger
Fraunhofer Institut für Kurzzeitdynamik (EMI)
Am Klingelberg 1
D - 79588 Efringen-Kirchen, Germany

ABSTRACT

A hierarchic order of influences on the wear resistance of ceramics is adopted here and applied in a modified form on recent ballistic results with sub-micrometer opaque and transparent oxide armor in order to understand interactions between different parameters that govern the ballistic strength. Previously contradictive experimental results such as a fine-grained transparent spinel that outperforms sapphire are explained when the dynamic stiffness (associated with Young's modulus) and the structure of ceramic fragmentation (on penetration) are positioned on the top of the hierarchy, whereas the macro-hardness is addressed as an important parameter, the influence of which, however, depends on the priority levels of the hierarchic ranking.

INTRODUCTION

Numerous studies have investigated influences of materials selection, microstructures and basic mechanical properties on the ballistic strength under different threats (e.g. AP or ball round ammunition) or with different incorporation of the ceramic into the composite armor system (e.g. with/without confinement or with different backing). Frequently, results reveal some major influences (e.g. hardness. Young's modulus or compressive strength) - which, however, turn out unimportant at changed configurations. On a general plane, the existing knowledge is limited to an extended body of individual results, but there is no real understanding of the reasons of contradictive facts.

A major shortcoming of many studies is their limitation to commercial grades of ceramic armor where the investigation of one influencing parameter (composition. hardness. ...) is obscured by variations of other properties (grain size. density. Young's modulus). However, there are also examples of investigations with laboratory-designed samples that did enable full control of all parameters but revealed, nevertheless, surprising results such as a higher ballistic strength of the armor with the *lower* hardness, *lower* Young's modulus, or *lower* compressive strength.

A similar inconsistency of results is known from *wear* investigations of brittle ceramics. These wear effects have been explained by a hierarchic model which places the micromechanical stability of grain boundaries (depending on grain boundary toughness and on subcritical crack growth resistance and affected by the stress transfer between the wear partners) on the top.[1] The frequently contradictive influence of the hardness is then explained as *depending on* hierarchically dominant parameters.

Now, flash X-ray investigations provided clear evidence that not only the metallic projectiles, but also the ceramic armor suffers substantial erosional wear during perforation.[2] Starting with a re-evaluation of exemplary previous and most recent ballistic data it is, therefore, the objective of the present work to modify the hierarchic wear model in a way that it can contribute to an advanced understanding of the ballistic performance of ceramic armor.

EVALUATION OF EXEMPLARY EXPERIMENTAL DATA

Evaluation of commercial ceramic armor

Fig. 1 illustrates results of a DoP (depth of penetration) investigation of 23 different alumina ceramics; ballistic trials of all the ceramics were performed using a tungsten long rod penetrator (172 grams, aspect ratio 14) with an impact velocity of 1335 m/s.[3] The mass efficiency E_m did not depend either upon the Hugoniot elastic limit or upon the spall strength, and Fig. 1 indicates that the mass efficiency was also independent of composition (additives/impurities), residual porosity, and grain size (with the grain size plot some benefit of *larger* grain sizes can be presumed but appears improbable from a ceramist's point of view).

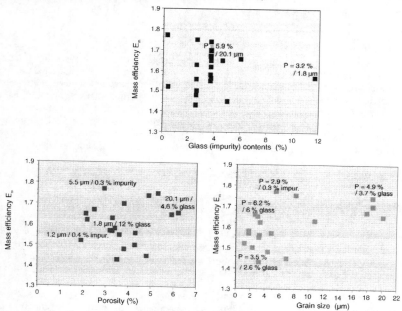

Fig. 1. Relationship of mass efficiency E_m (DoP tests with steel backing[3]) of alumina ceramics with microstructural data. Extreme values of the three parameters glass content, porosity and grain size are added to those data points they refer to.

At a first glance it is surprising that big microstructural and mechanical differences (up to 11.5 wt-% of additives/impurities, porosity 2-6 %, grain size 2-20 μm. Young's modulus 308-369

GPa) did *not* influence the ballistic performance in those tests. In fact, the extreme values of glass content, porosity and grain size in Fig. 1 indicate that the seemingly *missing* influence of these parameters may be a sum effect caused by the superposition of counteracting influences.

Evaluation of defined alumina grades in different ballistic tests

Similarly inconsistent results are also observed in some tests with armor samples where all material parameters, except one, are held constant:
- At high purity > 99.9% and density > 99% the hardness of sintered Al_2O_3 increases at small grain sizes,[4] and DoP investigations of the present authors with steel backing[5] have shown that generally the mass efficiency increases with the hardness[#] (provided all other parameters are constant). On the other hand, *at constant high hardness* (= small sub-μm grain size) the bending strength was reduced (by introducing flaws) by as much as 50 % *without* significant influence on the mass efficiency, and recent investigations revealed that even variations in the 4-point bending strength between 200 and 700 MPa did not affect the DoP result with steel backing (Fig. 2a).
- However, DoP tests with the same well-defined alumina ceramics with *aluminum* backing (AlCuMg1) and penetrated by 14.5 mm steel core (AP) projectiles revealed *no influence* of the hardness on the maximum mass efficiency (Fig. 2b).[6] Thus, seemingly missing influences are *not only* an issue of test conditions where several superposing effects may obscure results.

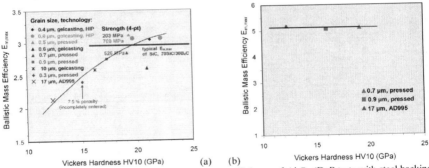

Fig. 2. (a) Influence of hardness on the maximum mass efficiency of Al_2O_3 (DoP tests with steel backing, tungsten rod penetrator with mass 44 g, aspect ratio 3.2, impact velocity 1250 m/s). (b) Missing influence of different ceramic grades in tests with 14.5 mm AP ammunition (1045 m/s) and Al backing.[6]

Ballistic strength of transparent armor: glass - sapphire - fine-grained ceramics (spinel. corundum)

A surprising result was observed in perforation tests of ceramic / glass (15-30 mm) / polycarbonate (4 mm) targets with 7.62 mm x 51 AP ammunition (v_p = 850 m/s) comparing the ballistic performance of different ceramic front tiles: sapphire ((0001) and ($11\bar{2}0$) orientations), transparent sintered fine-grained alumina (Fig. 3a)[7] and Mg-Al spinel (Fig. 3b).[8] In Fig. 4, the advantage of sintered Al_2O_3 over sapphire (similar performance of both orientations, eventually with slightly

---

[#] Hardness is always addressed here as a *macro*-hardness HV10 (Vickers hardness at 10 kgf testing load). With the strong indentation size effect in alumina, high hardness values can be measured using loads ≤ 500 g, but these data do not correlate well with wear or ballistic properties.

better performance of $(11\bar{2}0)$) can be explained by the higher hardness of the ceramic (HV10 ~ 21 GPa compared to 15-16 GPa for sapphire with these orientations). More difficult is an understanding of the *similar* ballistic strength of fine-grained spinel (HV10 ~ 15 GPa) and sub-μm Al_2O_3 regarding the big hardness difference of these ceramics. And, regarding lower spinel values of *all* basic data in Tab. I: what explains that the transparent spinel outperforms sapphire ?

The hierarchic order of influences introduced by the following section will try to answer these questions.

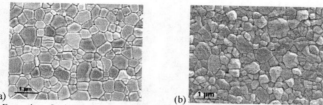

Fig. 3. Examples of microstructures of tested transparent (a) sintered sub-μm Al_2O_3 (av. grain size 0.55 μm) and (b) Mg-Al spinel (av. grain size 0.42 μm; similar ballistic results with 1.7 μm grain size).

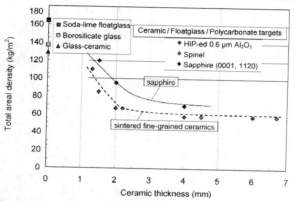

Fig. 4. Results of perforation tests (v_p = 850 m/s, AP ammunition). 1.5 mm thin front plies of sub-μm sintered Al_2O_3 reduce the normalized areal density by 50% (note, however: Al_2O_3 is not fully transparent at thickness > 1 mm![7]), and a similar stability is obtained with fine-grained sintered spinel that outperforms sapphire (and exhibits a real in-line transmission close to the theoretical limit[8]).

Table I. Basic mechanical data of sapphire, sintered transparent sub-μm Al_2O_3 and Mg-Al spinel.

		sapphire	sintered Al_2O_3	sintered Mg-Al spinel
Young's modulus	(GPa)	~ 400	~ 400	~ 275
Hardness HV10	(GPa)	15 - 16	20.5 - 21.5	14.5 - 15.0
4-point bending strength	(MPa)	400 - 600	600 - 700	200 - 250
Compressive strength	(MPa)	~ 2000	not determined	not determined
K_{IC}	(MPa√m)	2.0 - 2.8	~ 3.5	~ 1.8 - 2.2

HIERARCHY OF INFLUENCES ON CERAMIC WEAR MODIFIED TO DAMAGE OF CE-RAMIC ARMOR

Hierarchic ranking of influences and phases of projectile-target interaction

The hierarchic order of influences which govern the wear resistance of ceramics can be summarized as follows:[1]

❏ The *micromechanical stability of grain boundaries* occupies the upper rank of the hierarchy, it is governed by
- grain boundary strength (toughness),
- subcritical crack growth resistance, and
- stress transfer (← friction: surface roughness, lubricants).

The effects of all other influences depend on these upper rank parameters.

❏ Depending on grain boundary stability, a low or high frequency of local pull-out may induce opposite effects of microstructural or mechanical parameters: only with a high micromechanical stability of grain boundaries (→ few pull-out) a small grain size and high hardness will result in reduced wear whereas with a lower grain boundary stability (→ extended pull-out of individual crystals) a high hardness of the debris gives rise to more damage and increases wear.

When the hierarchy of *wear* influences shall be adopted for understanding the ballistic performance of ceramic armor, subcritical cracking and the grain boundary toughness appear negligible regarding an impact-induced damage propagation at sound velocity. The evaluation of other influences has to keep in mind that the energy of ceramic fragmentation consumes less than 1 % of the kinetic energy of a projectile[9,10] (i.e.: strength or K_{Ic} can hardly affect the ballistic strength directly), and that contributions to the total ballistic resistance come from different stages of the projectile-armor interaction where *different* properties of the ceramic need to be considered:

1. During a very short initial phase (< 10 µs), the high dynamic stiffness (determined by Young's modulus) overmatches the impact load and causes the nose of the projectile to dwell on the ceramic surface without penetrating it in the first moments. It is, therefore, expected that Young's modulus is most important for the contribution of this phase to the total ballistic strength of the ceramic armor.*

2. *Penetration* starts about 10-15 µs after the first contact (example: AP impact on B_4C[2]) with a relatively low penetration velocity, and it associates erosional wear and fragmentation of both the penetrator and the surrounding ceramic armor (where *the degree* of fragmentation can be rather different depending on the conditions). During this phase, the ceramic microstructure rapidly looses its cohesion strength, and it is assumed here that this extreme rate of destruction explains the observations of a missing influence of the static strength on the ballistic power of ceramics (e.g. Fig. 2a).

On the other hand, the deceleration of the projectile continues and the question arises *how* an almost instantly collapsing ceramic microstructure can slow down the projectile just in the moment of its own damage - an issue which will turn out most important *for the ranking* of the different ceramic properties that may affect the contribution of this phase to the total mass efficiency of the ceramic armor.

It is assumed here that the deceleration of the projectile during penetration (where the ceramic loses its cohesion strength during few µs) can be understood only if the *inertia* of the ceramic debris is large enough for a significant abrasive interaction with the penetrator. In fact, an order of magnitude approximation shows that with an average size of Al_2O_3 debris of 2.5 mm (measured by sieving[10]) its mass m = 0.045 g will exert a force F = m · a ≈ 1000 N when its acceleration at the ceramic-projectile interface is about (10-50) · 10^6 m/s^2 (this range is approximated

---

* Note that a high Young's modulus is generally associated with a high hardness of a crystal, but in the first phase (without penetration and without significant *plastic* deformation of the ceramic) its hardness (= resistance against micro*plastic* deformation [depending on the grain size]) is less important for the ballistic effect of the ceramic.

here from an evaluation of the penetration of AP projectiles into B₄C ceramics[11]). With this big force attributed by inertia to a 2.5 mm small fragment *after* complete loss of ceramic cohesion the ballistic effect of the ceramic during this phase of penetration will be governed by the following ranking of influences:

- the *mode of fragmentation* ($\rightarrow$ size and mass of debris $\Rightarrow$ *force* of inertia).
- the *hardness* of the ceramic debris ($\approx$ hardness of the ceramic microstructure).

As to the effect of the debris *size*, the force of inertia (F ~ m) *increases*, on the one hand, with the volume or the third power of the diameter whereas, on the other hand, the absolute number of ceramic-penetrator interactions *decreases* for a coarser mode of fragmentation. It can be shown that on summarizing these two effects the influence of the fragment size disappears for plane ceramic/projectile contact but increases strongly with the debris size for sharp point contacts (Appendix; this feature is, of course, bound to an upper size limit and becomes invalid when instead of common fragmentation the ceramic would fracture into just a few large pieces).

Static hardness vs Hugoniot elastic limit

Hardness is the resistance of the ceramic against localized *plastic* deformation. Regarding the dynamic character of the abrasive interaction the question arises whether in the present discussion the *static* hardness should be substituted by the dynamically measured Hugoniot elastic limit (HEL). Fig. 5 shows a close correlation of these two parameters covering *all* oxide and non-oxide armor ceramics with most different grain sizes, densities, and Young's moduli (only WC-based hard metals with a metallic binder phase are clearly outside of this correlation, the inclusion of *glass* into a general correlation depends on the assumption of a linear relationship). With this general correspondence of HEL and hardness it appears justified to discuss major influences on the ballistic strength on the basis of the more easily measured static macrohardness.

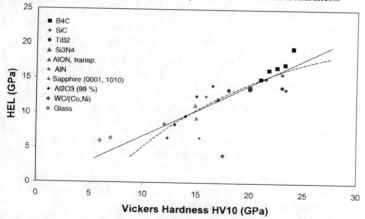

Fig. 5. Correlation of Hugoniot elastic limit (HEL) and macrohardness. For deriving this figure, HEL results from different references[12,13,14] (most of them without hardness information) were supplemented with hardness data HV10 from the present authors' experience. The solid line is a linear least-squares fit, the dotted fit results equally from logarithmic as from power approximations.

Dynamic vs static strength

With the limited present state of dynamic fracture mechanics it is difficult to say whether the DoP result of Fig. 2a ("*no influence* of strength differences of a factor of almost 4") is an indication of a general behavior or rather an individual case. Nevertheless, an advanced understanding is possible and can probably be obtained more easily when the time (or rate) effect is introduced into the fundamental expression

$$\sigma = E(D) \cdot \varepsilon$$

instead of starting from classical fracture mechanics of individual cracks (σ - stress, ε - strain, E - Young's modulus depending on damage D defined as the product of flaw frequency and flaw volume). Zhang and Hao [15] have used such an approach for describing the highly dynamic fracture of oil shale up to stressing rates of 10^4/s (which come close to the rates on ballistic impact). Their surprisingly simple model delivers expressions which enable important conclusions:

- At high stressing rates the fracture stress σ_{dyn} is described as the sum of two expressions: one is $\sigma_o(1-D)$ with the static strength σ_o, and the second is governed by $E_o(1-D)(d\varepsilon/dt)^{1/3}$. At a rate of 10^4/s, $\sigma_o(1-D)$ contributes to only about 3 % of σ_{dyn} of this material whereas 97 % of the high dynamic strength value are governed by the expression with the *static* Young's modulus E_o of the *un*damaged material.
  This result provides a more general frame for understanding the experimental finding of a missing influence of the static strength (4-pt bending) on the ballistic performance (Fig. 2a).
- An increase of the stressing rate from 10^2 to 10^4/s *decreases* the average fragment size (of oil shale) from 1-2 cm to about 0.8 mm - which would diminish the efficiency of an abrasive impact of ceramic fragments on a metallic penetrator. On the other hand, however, the calculation also shows how the fragment size can be *increased* at a given (high) stressing rate: (i) by high crack propagation velocity ($\rightarrow$ high Young's modulus) and (ii) by decreasing the number of cracks per volume (i.e. by larger crack spacing).
  For the development of future ceramic armor it appears, therefore, as an option to influence fragmentation by introducing defined populations of crack nuclei into the microstructures (while keeping E_o on a high level!).

From the above considerations it is concluded that the dynamic macroscopic failure on impact is more related to stressing rates and to the elastic response than to the static strength of ceramic armor components. The structure of fragmentation is a function of stressing rates (and modes) but depends also on E_o and on microstructural features.

Suggested hierarchic ranking of influences of basic ceramic properties on ballistic strength

Fig. 6 summarizes the considerations outlined above. Note that the *weight* of additive contributions from the dwell and the penetration phases to the total ballistic performance of a ceramic armor material will be rather different depending on the processes of projectile-armor interaction. Therefore, the high ranking of Young's modulus in the hierarchy of influences makes high values of this parameter generally beneficial but it is possible as well that under certain circumstances its influence can disappear and a ceramic grade with a high Young's modulus will not exhibit the best ballistic performance (e.g. under conditions where the contribution of dwell-phase interactions to the ballistic result is small).

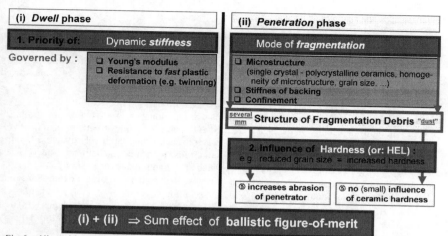

Fig. 6. Hierarchic order of parameters that influence the ballistic strength of ceramic armor.

When the *bending* (resp. *tensional*) strength was addressed above as a parameter which is probably less important for the ballistic performance of ceramic armor, the same should hold by similar reasons for the critical stress intensity K_{IC}. On the other hand, experimental investigations have revealed some correlation between the *compressive* strength of ceramics and DoP or v_c results (v_c - ballistic limit velocity on perforation),[12] and there is no doubt that the compressive strength is a valuable parameter when e.g. different ceramics shall be distinguished by a ranking of their quality.

Nevertheless, Fig. 6 does not incorporate the compressive strength because it is regarded as a mere composite parameter that is determined in a complicated way by stiffness ($\rightarrow$ Young's modulus), flaw population ($\rightarrow$ tensional strength), K_{IC} and other basic parameters. Therefore and regarding the general relationship between compressive deformation, Young's modulus and hardness it is assumed here that an additional introduction of the compressive strength into Fig. 6 would rather obscure than contribute to the physical figure of this hierarchy.

It is important to note that the concept presented here applies to the interaction of ceramics with small caliber projectiles at impact velocities up to about 1000 m/s. The influence of the different parameters in the case of long rod penetrators at high impact velocities is not clear yet.

IMPLICATIONS: EXAMPLES OF MODIFIED INTERPRETATION

With Fig. 6 it is easily understood why the substitution of a stiff steel backing (RHA) by a softer aluminum alloy was associated with a disappearance (or, at least, with a significant decrease) of the hardness influence displayed by Fig. 2: the softer backing gives rise to additional bending stresses and to a higher degree of ceramic fragmentation reducing the abrasive interaction with the penetrator.

A similar explanation holds for the surprising observation of Fig. 4 where the high ballistic strength of fine-grained transparent spinel outperforms sapphire (the latter exhibiting a higher level of *all* basic mechanical data - Tab. I). Experiments have shown that the effect of grain size on Hertzian contact damage in sintered alumina gives rise to an almost continuous transition between the performance of more fine-grained, coarser and single crystalline materials: the Normanski observation of crack networks in microstructures with average grain sizes of 3, 9, 15, 21, 35, and 48 μm revealed an increasing density of cracks (i.e. growing fragmentation) the coarser the microstructure was.[16] Within the framework of the hierarchy outlined by Fig. 6 this observation suggests on the one hand that *fine-grained* armor ceramics could profit from small grain sizes simply because of beneficial fragmentation (→ larger fragment size) and independent of the higher hardness such microstructures commonly exhibit. On the other hand, under localized load *single crystalline* alumina (sapphire) exhibits a more brittle behavior than sintered Al_2O_3 ceramics - a difference that is well known from indentation experiments and is caused by both the low K_{IC} of sapphire (Tab. I) and by the ease of straight crack propagation there. Thus, it is expected that a closer network of more cracks (→ smaller fragments) is generated upon first contact of the projectile with the single crystalline material which is damaged in a way that, finally, its ballistic strength is lower than anticipated from its hardness (and lower than that of the fine-grained polycrystalline transparent spinel). Further, twinning is commonly rather active in single crystals and in very coarse microstructures where this process is not restricted by a close network of grain boundaries. Therefore, some fast plastic deformation may decrease the dynamic stiffness of sapphire during the dwell phase of impact whereas this process is highly improbable in the fine-grained sintered spinel and corundum ceramics of the test of Fig. 4.

Woodward et al.[9] gave an instructive example of the role of a confinement within the frame of the hierarchy of Fig. 6. These investigations were performed with a similar aluminum backing as used in some tests of Strassburger et al.[6] but had addressed alumina ceramics with different compositions and, therefore, with different values of Young's modulus. As to be expected with the present discussion, the experiments *with confinement* showed a lower degree of ceramic fragmentation: *less* very small fragments < 0.18 mm and an increased quantity of larger, relevant for the ballistic strength pieces with a size of several millimeters. For the alumina grades AD90/AD96/AD995 this beneficial fragmentation was associated with a smaller "confined" depth of penetration (48 mm) than observed in the tests without confinement (55 mm). In agreement with Fig. 6 above these investigations also show a correlation of *smaller DoP* values with *higher hardness* of the ceramic armor *for the confined tests only* whereas the hardness was unimportant for the results obtained without confinement (where a higher amount of very small fragments was observed).

CONCLUSIONS

From a critical analysis of previous and recent experimental ballistic investigations it is suggested that, in analogy with wear experiences, the multitude of microstructural and mechanical parameters of ceramic armor materials (porosity, grain size, Young's modulus, HEL, hardness, strength, K_{Ic}, ...) do not form an uncontrollable n-dimensional network of arbitrary influences but govern the ballistic strength according to a clear hierarchic order.

The way one of these influences depends on processes of higher hierarchic ranking is similar as known from wear mechanisms. A specific structure of the hierarchy of influences which govern the ballistic strength was proposed here considering contributions from the dwell and the penetration phases of projectile-armor interaction. Apparently, this hierarchic order of influences can explain previously contradictive ballistic results.

APPENDIX
Size and Configuration Influences on the Effect of Ceramic Fragments on a Metallic Penetrator

When the ceramic fragment is approximated as a small cube with edge length d, two extreme configurations can be addressed as a frame for a simplified understanding of different possible interactions.

(i) When one of the flat square faces of the cube is pressed to a flat metallic surface by a force F, the resulting pressure P is F/d^2, and the wear effect caused by this one ceramic particle will be related (probably proportionally) to this pressure. Coarser fragmentation with an increase of the particle size $d_2 = k \cdot d_1$ will increase the force of inertia F (F = m·a, with mass m ~ d^3) by k^3 and the area by k^2, and the resulting pressure will increase as $P_2 = k \cdot P_1$. However, the total number n of ceramic fragments *decreases* as $n_2 \approx n_1/k$ when the ceramic is fragmented into *larger* pieces.[&] Thus, the sum effect of the increased pressure per particle and of their reduced number is expected to remain nearly constant independent of the fragment size.

(ii) Much more probable is a configuration where a sharp ceramic fragment contacts the flat surface of the penetrator by one of its edges or corners. The case is similar to the configuration on hardness testing with sharp indenters, and with a force F the size L of the indent will be related to the hardness H_p of the projectile as $F/L^2 \sim H_p$. Therefore, $L \sim F^{1/2}$. Because of the proportionality of L and the total diameter D_{pl} of the plastically deformed zone (generated in the metal under a ceramic tip), $D_{pl} \sim L$ gives $D_{pl} \sim F^{1/2}$, and the volume of the deformed zone will be $V_{pl} \sim F^{3/2}$. Again, an increase of the particle size by a factor k will increase the force of inertia by k^3 which increases the deformed volume V_{pl} under the ceramic tip as $V_{pl,2} = k^{9/2} \cdot V_{pl,2}$, but again the number of contacts will decrease as $n_2 \approx n_1/k$. Thus, assuming the energy introduced into the metallic body by this interaction as proportional to the sum of all V_{pl} the net effect of an increased fragment size should be proportional to $k^{7/2}$ in this configuration.

Summarizing over (i) and (ii) and regarding the different probabilities of these configurations it is expected that the ballistic strength of a ceramic tile increases with the debris size (limitation of this model: the number of fragments must not be too small).

ACKNOWLEDGEMENTS

The present work uses resulted from investigations at IKTS Dresden and at EMI Freiburg that were performed with financial support of the German Ministry of Defense under contracts E/E210/0D008/ 11452, E/E91S/3A621/T5185, and E/E91S/4A299/3F034. Other parts of this work were funded by an internal project "MMM Tools: Multiscale Materials Modeling" of the Fraunhofer Society.

REFERENCES

[1] A. Krell, "Improved Hardness and Hierarchic Influences on Wear in Submicron Sintered Alumina," *Mater. Sci. Eng. A* **209** [1-2] 156-63 (1996).

[2] W.A. Gooch, M.S. Burkins, P.Kingman, G.Hauver, P.Netherwood, and R Benck, "Dynamic X-Ray Imaging of 7.62-mm APM2 Projectiles Penetrating Boron Carbide," pp. 901-908 in: Proc. of the *18th Int. Symp. on Ballistics*, Vol. 2, Technomic Publishing Co., Lancaster/PA, 1999.

[3] B. James, "The Influence of the Material Properties of Alumina on Ballistic Performance," pp. 3-9 in: *Proc. of the 15th Int. Symp. on Ballistics*, May 21-24, Jerusalem, Israel, 1995 (ISBN 0-961-8156-0-4).

[&] Approximation valid for larger values of n_1, n_2; the exact solution depends on the configuration, for fragments with circular cross sections circumferencing a cylindrical penetrator it is e.g. $n_2 = n_1/k + \pi(k-1)/k$.

[4] A. Krell, "A New Look at the Influences of Load, Grain Size, and Grain Boundaries on the Room temperature Hardness of Ceramics," *Int. J. Refract. Metals & Hard Mat.* **16** [4-6] 331-335 (1998).

[5] A. Krell and E. Strassburger, "High-Purity Submicron α-Al_2O_3 Armor Ceramics: Design, Manufacture, and Ballistic Performance," *Ceram. Trans.* **134** [1-2] 463-471 (2001).

[6] E. Strassburger, B. Lexow, and A. Krell, „Ceramic Armor with Submicron Alumina against AP Projectiles," *Ceram. Trans.* **134** [1-2] 83-90 (2001).

[7] A. Krell, G. Baur, and C. Daehne, "Transparent Sintered Sub-μm Al_2O_3 with IR Transmissivity Equal to Sapphire," pp. 100-207 in: R.W. Tustison (ed.), *Window and Dome Technologies VIII, Proc. of SPIE conference* (Orlando/FL, April 22-23, 2003), Vol. 5078, Washington, 2003.

[8] A. Krell, T. Hutzler, and J. Klimke, "Physics and Technology of Transparent Ceramic Armor: Sintered Al_2O_3 vs Cubic Materials," Paper 14 in: *Proc. Specialists Meeting on "Nanomaterials Technology for Military Vehicle Structural Applications"* (NATO Research and Technology Organization [RTO)], Applied Vehicle Technological Panel [AVT]), Granada/Spain, Oct. 3-4 2005.

[9] R.L.Woodward, G.A. Gooch Jr., R.G. O'Donnell, W.J. Perciballi, B.J. Baxter, and S.D. Pattie, "A Study of Fragmentation in the Ballistic Impact of Ceramics," *Int. J. Impact Engn.* **15** [5] 605-618 (1994).

[10] E. Wollmann, E. Lach, and B. Wellige, "Investigation of the Ballistic Strength of Ceramic Inserts Against Kinetic Projectiles," German Research Report RT 513/95, German-French Research Institute Saint Louis (ISL), 1995.

[11] C.A. Anderson, Jr, M.S. Burkins, J.D. Walker, W.A. Gooch, "Time-Resolved Penetration of B_4C Tiles by the APM2 Bullet," Computer Modeling in Eng. Sci. **8** [2] 91-104 (2005).

[12] A.M. Rajendran, T.J. Holmquist, D.W. Templeton, and K.D. Bishnoi, "A Ceramic Armor Material Database," U.S. Army TARDEC, Detroit Arsenal, Warren/MI, 1999.

[13] J. Hyun, S.M. Sharma, and Y.M. Gupta, "Ruby R-line Shifts for Shock Compression Along (1120)," J. Appl. Phys. **84** [4] 1947-1952 (1998).

[14] J. Cazamias, S.J. Fiske, and S. Bless, "The Hugoniot Elastic Limit of ALON," *AIP Conference Proc.* **620** [1] 767-770 (2001).

[15] Y.-Q. Zhang, H. Hao, "Dynamic Fracture in Brittle Solids at High Rates of Loading," J. Appl. Mech. (Trans. of ASME) **70** [5] 454-457 (2003).

[16] F. Guiberteau, N.P. Padture, and B.R. Lawn, "Effect of Grain Size on Hertzian Contact Damage in Alumina," *J. Am. Ceram. Soc.* **77** [7] 1825-1831 (1994).

CONCEPTS FOR ENERGY ABSORPTION AND DISSIPATION IN CERAMIC ARMOR

Dong-Kyu Kim[1] *, Jonathan Bell[1] *, Dechang Jia[2] *, Waltraud M. Kriven[1] *, ** and Victor Kelsey[3] *

[1] Department of Materials Science and Engineering, University of Illinois at Urbana-Champaign, Urbana, IL, 61801, USA, [2] Department of Materials Science and Engineering, Harbin Institute of Technology, Harbin, 150001, China, [3] Armor Holdings, Aerospace and Defense Group, Phoenix, AZ 85044, USA

ABSTRACT

As possible new ceramic armor materials, different kinds of composites having potential different energy absorption mechanisms were fabricated and their ballistic performances were qualitatively tested under unconstrained conditions. A 50 vol% dicalcium silicate ($2CaO \bullet SiO_2$, or "C_2S") – 50 vol% calcium zirconate ($CaO \bullet ZrO_2$, or "CZ") composite, a 25 vol% Al_2O_3 - 25 vol% $NiAl_2O_4$ - 25 vol% TiC - 25 vol% ZrO_2 quadruplex composite, a similar nano grain-sized, four-phase (quadruplex) composite of the same composition, and a 50 vol% Al_2O_3 coated – 50 vol% TiB_2 platelet composite were fabricated by hot pressing. A multilayer, carbon fiber-reinforced, nanoporous, geopolymer composite of approximate composition ($K_2O \bullet Al_2O_3 \bullet 4SiO_2 \bullet 12H_2O$) was also fabricated as a possible light-weight armor material.

The quadruplex composite hot pressed at 1600°C for 1 h at the pressure of 68 MPa in the Ar atmosphere had nearly a full density and possessed hardness, 3-point bend strength and toughness values of 1251 Kg/mm², 984 MPa and 4.74 MPa•m$^{1/2}$, respectively. The results of qualitative ballistic performance testing indicated that the four-phase composite, nano-sized quadruplex composite, and 50 vol% Al_2O_3 coated - 50 vol% TiB_2 platelet composite had ballistic performances comparable to that of standard Al_2O_3.

INTRODUCTION

In a typical composite armor system, ceramic plates and layers of fabric are bonded together with a polymeric binder.[1] The ceramic plate absorbs the major part of the kinetic energy of a projectile and decelerates it. Silicon carbide (SiC), aluminum oxide (Al_2O_3), aluminum oxynitride (AlON), boron carbide (B_4C), and silicon nitride (Si_3N_4) are typical ceramics used in ceramic armor systems. Despite its excellent ballistic performance and having the lowest density, the high cost and complexity in processing of B_4C are major issues for any actual application of this material.[2]

Transformation toughening in zirconia-toughened ceramics (ZTC) occurs because of the stress-induced, tetragonal to monoclinic transformation in zirconia.[3] The metastable, tetragonal phases transform into stable, monoclinic structures in the deviatoric stress field of a propagating crack, and alter the crack tip stress field, dissipating the crack energy and thereby finally increasing the toughness of the composite. This transformation in zirconia involves a dilatational strain of 4.9% at room temperature and a shear strain of about 16%.[4, 5] Dicalcium silicate ($2CaO \bullet SiO_2$, C_2S) has a +12 vol% expansion at 490°C on cooling.[5-11] Because of this volumetric change, this material may be able to produce a stress field which can be utilized as a energy absorbing and toughening source. Calcium zirconate ($CaO \bullet ZrO_2$ or "CZ") is chemically compatible with C_2S and used as a matrix phase in the C_2S-CZ composite. Multiphase composites have minimized grain sizes due to retarded grain growth caused by extended diffusion distances in the microstructure.[12,13] Smaller grain sizes produce better ballistic performances due to enhanced hardness.[14] The TiB_2-Al_2O_3 composite has been studied as a low cost armor system.[15,16] Geopolymer is an amorphous aluminosilicate structure formed by geopolymerization between alumino-silicate oxides and alkali metasilicate solution and may find use as refractory adhesives or as light-weight, advanced composite materials.[17-29]

In this research, five different kinds of ceramics having potentially different energy absorbing mechanisms were manufactured and tested to investigate their performances as armor materials. A 50 vol% C_2S – 50 vol% CZ composite, a 25 vol% Al_2O_3 – 25 vol% $NiAl_2O_4$ – 25 vol% TiC – 25 vol% ZrO_2 composite, a nano grain-sized quadruplex composite, 50 vol% Al_2O_3 coated 50 vol% TiB_2 platelet composite, and a multilayer carbon fiber reinforced geopolymer composite were fabricated, and their ballistic performances were tested by measuring the depths of penetration.[30,31]

EXPERIMENTAL PROCEDURES

1.   Processing
1.1 C_2S-CZ composite

C_2S and CZ composite powder was chemically synthesized by the organic, steric entrapment method [32, 33] as described in Fig. 1. Calcium nitrate tetrahydrate [$Ca(NO_3)_2$ • 4 H_2O, 99%, Aldrich Chemical Inc., WI, USA] and LUDOX SK [Dupont Chemical Inc., 25 wt% colloidal silica suspension in water] were used as Ca^{+2} and Si^{+4} sources, respectively, for the synthesis of C_2S. To synthesize CZ powder, calcium nitrate tetrahydrate and zirconium (IV) propoxide [70 wt% solution in 1-propanol, Aldrich Chemical Inc.] were used as Ca^{+2} and Zr^{+4} sources, respectively. As polymeric, ion carriers, PVA [polyvinyl alcohol, 205S, Celanese Ltd., Dallas, TX, USA] and polyethylene glycol [PEG, MW=200, Aldrich Chemical Inc.] were used for the synthesis of C_2S and CZ, respectively. Chemicals were mixed either in the deionized water or in ethanol [ethyl alcohol USP, AAPER ALCOL and Chemical, Shelbyville, KY, USA] for 20 min, for the synthesis of C_2S and CZ, respectively. A polymeric source was added to the solution, and then followed by another 50 min of mixing for homogenization of solution. The final mixture was heated until most of the water was removed and then dried overnight in the oven at 150°C. The dried cakes were pulverized and calcined at 800°C for 1 h. The calcined powder was attrition milled for 1 h and sieved through a 100 mesh sieve. The final powder was hot pressed into 7.62 cm diameter disc sample at 1400°C for 2 h under a pressure of 31 MPa in Ar atmosphere.

1.2   Multiphase composites

A 50 vol% Al_2O_3 – 50 vol% $NiAl_2O_4$, two-phase (duplex), ceramic composite powder was chemically fabricated by the organic, steric entrapment method.[8] Aluminum nitrate nonahydrate [$Al(NO_3)_3$ • 9H_2O, 98+%, Aldrich Chemical Inc., Milwaukee, WI, USA] and nickel hexanitrate [$Ni(NO_3)_2$•6H_2O, Aldrich Chemical Inc.] were used as chemical sources for Al^{+3} and Ni^{+2}, respectively. Appropriate amounts of 50 vol% Al_2O_3 - 50 vol% $NiAl_2O_4$ *in situ* composite powder were mixed with given amounts of 3 mole% yttria (Y_2O_3) stabilized tetragonal zirconia polycrystals [3Y–TZP, Tosoh chemicals Inc., Tokyo, Japan] and TiC [99.5 %, Alfa Aesar, Word Hill, MA, USA] powder, by ball milling for 24 h to make a four-phase (quadruplex), 25 vol% Al_2O_3 - 25 vol% $NiAl_2O_4$ - 25 vol% TiC - 25 vol% 3Y-TZP composite powder. Nano-sized quadruplex composite powder was obtained by dry milling of the 4-phase powder using a planetary micromill (Model Pulverisette 7, Fritsch Inc., Idar-Oberstein, Germany) for 1 h at 700 rpm. ZrO_2 balls of 5 and 8 mm diameter were used. The planetary milled powder formed soft agglomerates of fine particles having an average size about 50 nm, as shown in Fig. 2. The four-phase and nano-sized, four-phase, composite powder were hot pressed at 1600°C for 2 h, at a pressure of 31 MPa in Ar atmosphere.

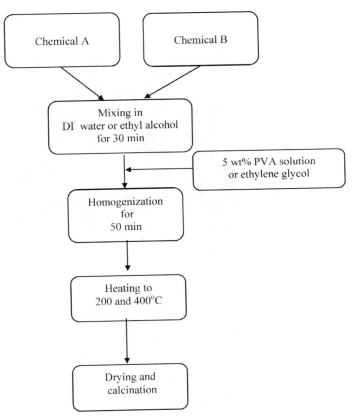

Fig. 1. Schematic diagram of the organic, steric entrapment method for synthesis for oxide ceramic powders.

1.3 Al_2O_3 coated TiB_2 platelet composite

Titanium diboride (TiB_2) platelets [HCT-30, General Electric Ceramics, Cleveland, OH] having an average particle size of 14 μm were coated with Al_2O_3 using the organic steric entrapment method, as shown in Fig. 3, to produce 50 vol% Al_2O_3 coated - 50 vol% TiB_2 platelets. Appropriate amounts of aluminum nitrate nonahydrate and PVA [Aldrich Chemical

Inc., molecular weight = 9500] were mixed in DI water for 30 min. Suitable amounts of mixed solution was diluted with extra DI water to produce desired amounts of Al_2O_3 coating on the platelets. The diluted solution was mixed with TiB_2 platelets. Ultrasonication for 1.5 hours was applied to improve the coating of the platelets. All liquid was heated out of the final solution until only powder remained. The powder was then gently ground and sieved through a 450 μm sieve. After sieving, the powder was calcinated at 450°C for one hour. The calcined powder was hot pressed at 1650°C for 2 h under a pressure of 31 MPa, in an Ar atmosphere, into a disc of 7.62 cm diameter.

500 nm

Fig. 2. SEM micrograph showing formation of nano powders of the 25 vol% Al_2O_3 –25 vol% $NiAl_2O_4$ - 25 vol% TiC - 25 vol% 3Y-TZP four-phase composite after 1 h of dry planetary milling.

1.4   Carbon fiber - geopolymer composite

4 SiO_2 : Al_2O_3 : K_2O : 12 H_2O geopolymer was made by mixing highly reactive metakaolin [Metamax high reactivity metakaolin (HRM), Engelhard corporation, Iselin, NJ USA] with potassium metasilicate solution. Potassium hydroxide [KOH, Fisher scientific, Suwanee, GA USA] was added to distilled water and then fumed silica [Carbo-sil EH-5, Cabot corp. T. H. Hilson company, Wheaton, IL USA] was added to make potassium metasilicate. Each of 27 carbon fiber [plain weave carbon fiber fabric, 3K, Fiber Glast Developments Corporation, Brookville OH USA] layers was hand infiltrated with geopolymer and laid into the mold one by one to make a composite. The resulting composite was cured at 60°C for 24 h under 12 MPa of pressure. The cured composite was then removed from the mold and machined into a 7.62 cm diameter by 2.8 cm high cylindrical sample.

2.   Characterization

The microstructures of the sintered ceramics were examined by scanning electron microscopy [SEM, Model S-4700, Hitachi, Osaka, Japan].

Flexural strengths were measured with a screw-driven machine [Model 4502, Instron Corp., Canton, MA] in 3–point bend testing, where 3 to 5 samples were tested for a given condition. The specimen size was 3 mm (H) x 4 mm (W) x 40 mm (L), the supporting span was

30 mm, and the cross head speed was 0.1 mm/min. Vickers indentation hardness was determined using a commercial microhardness tester [Zwick 3212 microhardness tester, Mark V Laboratory Inc., East Granby, CT]. The surface of sintered pellets was micropolished down to 1 μm with diamond paste. 0.8 kg of low load was applied to avoid possible crack formation during indentation. 3 mm/min of impact rate, and 10 sec of loading time were applied. Toughness values were measured by the Vickers and Knoop indentation methods [32]. The loads used for the Knoop and Vickers indentations to measure a toughness value were 2 kg and 8 kg, respectively. At least 15 measurements were made to determine the average hardness and toughness values.

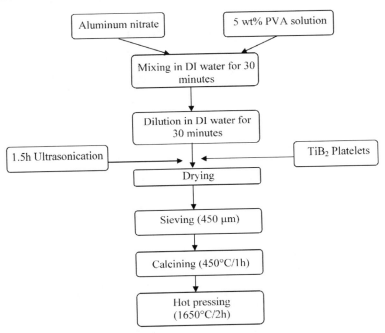

Fig. 3. Schematic diagram showing the procedures for coating Al_2O_3 on TiB_2 platelets using the organic steric entrapment method.

The ballistic performance of different unconstrained ceramic composites was tested qualitatively by measuring the depth of penetration. The schematic diagram of the target assembly is shown in Fig. 4. The diameters of all samples were 7.62 cm. The final thicknesses of samples were different for each composite.

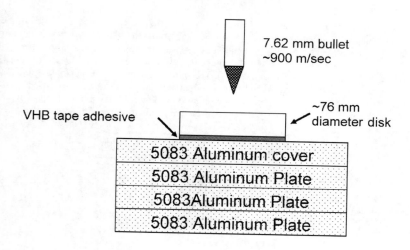

Fig. 4. Schematic diagram of the target assembly for the ballistic evaluation of unconstrained ceramic composites.

RESULTS AND DISCUSSION

The SEM micrograph of 50 vol% C_2S - 50 vol% CZ composite hot pressed at 1400°C for 2 h at a pressure of 31 MPa is shown in Fig. 5. Despite the homogeneous nature of microstructure, the composites still had some pores after hot pressing. The hot pressed C_2S-CZ composite possessed a relatively low hardness of 246 Kg/mm^2.

The quadruplex composite hot pressed at 1600°C for 1 h at a pressure of 68 MPa in an Ar atmosphere had nearly full density without any indication of pores, as shown in Fig. 6. The hardness, 3-point bend strength and toughness value of the composite were 1251 Kg/mm^2, 984 ± 29 MPa and 4.74 MPa•m$^{1/2}$, respectively. The excellent mechanical properties of this material are attributed to the retarded grain growth due to the extended diffusion distance among same grains separated by same-phase grains in the multi-phase composite.

The SEM micrographs of as-received TiB_2 platelets and 15 vol% Al_2O_3 coated 85 vol% TiB_2 platelets are shown in Fig. 7. The SEM micrograph of Fig. 7 (a) shows that the as-received TiB_2 platelets had an average diameter of 14 µm. Fig. 7 (b) and (c) shows homogeneously coated TiB_2 platelets with Al_2O_3. The calcined, Al_2O_3 coated, TiB_2 platelets showed drying cracks around the platelets in the as-deposit powder. The Al_2O_3 coating material was composed of nano-sized particles as can be seen in Fig. 7 (d). The SEM micrographs of 50 vol% Al_2O_3 coated 50 vol% TiB_2 platelet composite hot pressed at 1650°C for 2 h are shown in Fig. 8. Except for local formation of pores due to misalignments among TiB_2 platelets, the composite had a relatively dense microstructure.

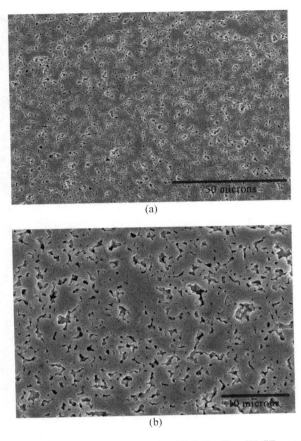

(a)

(b)

Fig. 5. SEM micrographs of microstructures of 50 vol% C₂S – 50 vol% CZ composite hot pressed at 1400°C for 2 h under a pressure of 31 MPa in an Ar atmosphere.

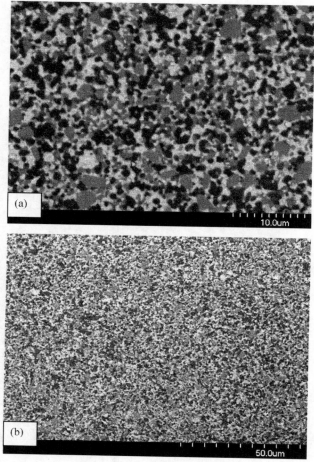

Fig. 6. SEM micrographs of microstructures of quadruplex, 25 vol% Al_2O_3 - 25 vol% $NiAl_2O_4$ - 25 vol% TiC - 25 vol% ZrO_2 composite, hot pressed at 1600°C for 1 h under a pressure of 68 MPa in an Ar atmosphere.

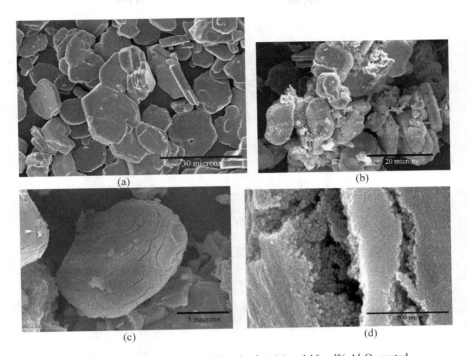

Figure 7. SEM micrographs of uncoated TiB$_2$ platelets (a) and 15 vol% Al$_2$O$_3$ coated – 85 vol% TiB$_2$ platelets (b), (c) and (d).

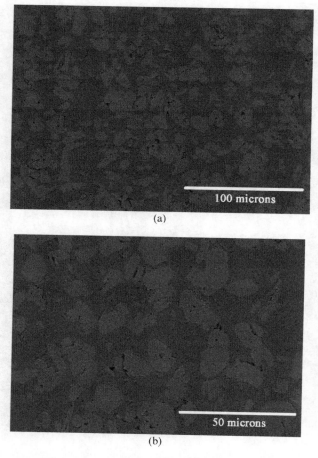

(a)

(b)

Fig. 8. SEM micrographs of microstructures of 50 vol% Al_2O_3 coated 50 vol% TiB_2 platelet composite hot pressed at 1650°C for 2 h with the pressure of 31 MPa in an Ar atmosphere.

The results of ballistic performance testing of different ceramic composites are summarized in Table I. AD-90 Al_2O_3 was used as a standard material and its depth of penetration was compared to those of other ceramic composites. The depths of penetration of the 25 vol% Al_2O_3 - 25 vol% $NiAl_2O_4$ - 25 vol% TiC - 25 vol% ZrO_2 composite, the nano-sized four-phase composite of the same composition, and the 50 vol% Al_2O_3 coated 50 vol% TiB_2 composite were similar to that of standard material. The 50 vol% C_2S – 50 vol% CZ composite and carbon fiber reinforced geopolymer composite allowed a much higher depth of penetration compared to that of AD-90 Al_2O_3. Despite the very high volumetric expansion of +12% upon phase transformation of CZ, the relatively lower hardness of 246 Kg/mm^2 was attributed to relatively poor ballistic performance of this material. It is anticipated that since the 50 vol% C_2S – 50 vol% CZ composite was not tested in a confining frame, the C_2S transformation was not nucleated, and the sample simply fractured instead. Similarly, the relatively soft nature of the carbon fiber and geopolymer also is thought to be the main reason for the relatively poor ballistic performance in the combined composite. Because of good performance observed in the three other composites, four-phase, nano-sized quadruplex, and Al_2O_3-TiB_2 platelet composite, which were at least as good as that of standard AD-90 Al_2O_3, these composites need to be further quantitatively re-evaluated for their ballistic performance, particularly in a confining frame.

Table I. Results of ballistic performance testing of unconstrained ceramic composites

Target material	Thickness (mm)	Velocity (m/sec)	Depth of penetration (mm)	Comments
AD-90 Alumina	9.53	913	~ 2.5 (dent)	Standard material
50 vol% CZ – 50 vol% C_2S	12.45	913	> 38	-
Four-phase composite	12.15	910	~ 2.5 (dent)	-
Nano-sized quadruplex composite	12.10	909	~ 2.5 (dent)	-
50 vol% Al_2O_3 coated 50 vol% TiB_2 platelets	14.10	915	~ 2.5 (dent)	-
Carbon fiber reinforced geopolymer composite	-	911	> 38	-

SUMMARY

Various ceramic composites having different potential energy-absorbing mechanisms were fabricated and tested for their unconstrained ballistic performance, as possible new armor materials.

Despite the high potential volume change on phase transformation in dicalcium silicate (C_2S), the C_2S-CZ composite showed relatively poor ballistic performance. This was thought to be due to the transformation not having been nucleated and also due to its lower hardness of 246 Kg/mm^2. Because of the soft nature of carbon fiber reinforced geopolymer, this composite also allowed relatively easy penetration by the ballistic projectile.

A quadruplex 25 vol% Al_2O_3 - 25 vol% $NiAl_2O_4$ - 25 vol% TiC - 25 vol% ZrO_2 composite hot pressed at 1600°C for 1 h, became nearly fully dense, and possessed hardness, 3-point bend strength and toughness values of 1251 Kg/mm^2, 984 ± 29 MPa and 4.74 MPa•$m^{1/2}$, respectively. A duplex 50 vol% Al_2O_3 coated 50 vol% TiB_2 composite was hot pressed to a relatively dense microstructure except for the formation of local pores due to misalignments among platelets. The quadruplex composite, nano-sized four-phase composite and 50 vol% Al_2O_3 coated 50 vol% TiB_2 composite showed comparable ballistic performance to that of standard Al_2O_3. Hence, more detailed ballistic performances of these materials particularly under confining conditions where a phase transformation is more likely to be nucleated, need to be further evaluated.

ACKNOWLEDGEMENTS

Part of this work was carried out in the Center for Microanalysis of Materials, in the Frederick Seitz Materials Research Laboratory, which is partially supported by the U.S. Department of Energy under grant DEFG02-91-ER45439. This work was partially supported by the US Army TARDEC division under GSA Contract number W56HZV-05-P-L681. It was also partially supported by a research grant from the GE Corporate Research and Development Department of the General Electric Company, Niskayuna, NY, USA.

REFERENCES

[1] A. V. Gorik, O. N. Griror'ev, D. Yu., Ostrovoi, V. G. Piskunov and V. N. Cherednikov, "Experimental and Theoretical Studies on the Nonlinear Stress-Strain State of a Laminated Ceramic Composite," *Strength of Materials*, **33** 526-534 (2001).

[2] L. R. Vyshnyakov, A. V. Mazna, A. V. Neshpor, V. A. Kokhanyi and O. N. Oleksyuk, "Influence of Structural and Technological Factors on the Efficiency of Armor Elements Based on Ceramics," *Strength of Materials*, **36** 643-648 (2004).

[3] R. C. Garvie, R. H. J. Hannink and R. T. Pascoe, "Ceramic Steel?," *Nature*, **258** 703 – 704 (1975).

[4] W. M. Kriven, "Martensite Theory and Twinning in Composite Zirconia Ceramics," pp. 168-183, In Advances in Ceramics, Vol. **3**, Science and Technology of Zirconia. Edited by A. H. Heuer and L. W. Hobbs. The American Ceramic Society, Columbus, OH, 1981.

[5] W. M. Kriven, "The Transformation Mechanism of Spherical Zirconia Particles in Alumina," *Advances in Ceramics*, **12** 64-77 (1984).

[6] W. M. Kriven, "Possible Alternative Transformation Tougheners to Zirconia: Crystallographic Aspects," *J. Am. Ceram. Soc.* **71** [12] 1021-1030 (1988).

[7] W. M. Kriven, C. J. Chan and E. A. Barinek, "Particle Size Effect of Dicalcium Silicate in a Calcium Zirconate Matrix", *Advances in Ceramics*, **24A** 145-155 (1988).

[8] I. Nettleship, K. G. Slavick, Y. J. Kim and W. M. Kriven, "Phase Transformations in Dicalcium Silicate. I: Fabrication and Phase Stability," *J. Am. Ceram. Soc.*, **75** [9] 2400-2406 (1992).

[9] Y. J. Kim, I. Nettleship and W. M. Kriven, "Phase Transformation in Dicalcium Silicate, II : TEM Studies of Crystallography, Microstructures and Mechanisms," *J. Am. Ceram. Soc.*, **75** [9] 2407-2419 (1992).

[10] I. Nettleship, K. G. Slavick, Y. J. Kim and W. M. Kriven, "Phase Transformations in Dicalcium Silicate. III: Effects of Barium on the Stability of Fine-grained $\alpha'[L]$ and β Phases," *J. Am. Ceram. Soc.*, **76** [10] 2628-2634 (1993).

[11] T. I. Hou and W. M. Kriven, "Mechanical Properties and Microstructure of $CaZrO_3$-Ca_2SiO_4 Composites," *J. Am. Ceram. Soc.*, **77** [1] 65-72 (1994).

[12] J. D. French, M. P. Harmer, H. M. Chan, and G. A. Miller, "Coarsening Resistant Dual-Phase Interpenetrating Microstructures," *J. Am. Ceram. Soc.*, **73** [8] 2508-10 (1990).

[13] D. K. Kim and W. M. Kriven, "Processing and Characterization of Multi-Phase Ceramic Composites, Part I : Duplex Composites Formed *In-Situ*," *J. Am. Ceramic. Soc.*, (2007), in press.

[14] A. Krell and P. Blank, "Grain Size Dependence of Hardness in Dense Submicrometer Alumina," *J. Am. Ceram. Soc.*, **78** [14] 1118-1120 (1995).

[15] G. A. Gilde, J. W. Adams, M. Burkins, M. Motyka, P. J. Patel, E. Chin, L. Prokurat Franks, M. P. Sutaria and M. Rigali, "Processing of Al_2O_3/TiB_2 Composites for Penetration Resistance," *Cer. Eng. Sci. Proc.*, **22** 331-342 (2001).

[16] G. A. Gilde and J. W. Adams, "Processing and Ballistic Performance of Al_2O_3/TiB_2 Composites," *Cer. Eng. Sci. Proc.*, **26** 257-262 (2005).

[17] J. Davidovits, "Geopolymers : Inorganic Polymeric New Materials," *J. Therm. Anal. (UK)*, 37 [8] 1633-1656 (1991).

[18] C. G. Papakonstantinou, P. Balaguru and R. E. Lyon, "Comparative Study of High Temperature Composites," *Composites : Part B*, **32** [8] 637-649 (2001).

[19] W. M. Kriven, J. L. Bell and M.Gordon, "Microstructure and Microchemistry of Fully-Reacted Geopolymers and Geopolymer Matrix Composites," *Ceramic Transactions* vol. **153**, 227-250 (2003).

[20] D. C. Comrie and W. M. Kriven, "Composite Cold Ceramic Geopolymer in a Refractory Application," *Ceramic Transactions* vol. **153**, 211-225 (2003).

[21] W. M. Kriven and J. L. Bell, "Effect of Alkali Choice on Geopolymer Properties," *Cer. Eng. and Sci. Proc.* vol. **25** [3-4] 99-104 (2004).

[22] W. M. Kriven, J. L. Bell and M. Gordon, "Geopolymer Refractories for the Glass Manufacturing Industry," *Cer. Eng. and Sci. Proc.* vol. **25** [1] 57-79 (2004).

[23] M. Gordon, J. Bell and W. M. Kriven, "Comparison of Naturally and Synthetically-Derived, Potassium-Based Geopolymers," *Ceramic Transactions,* vol. **165**. Advances in Ceramic Matrix Composites **X**. Edited by J. P. Singh, N. P. Bansal and W. M. Kriven 95-106 (2005).

[24] P. Duxson, G. C. Lukey, J. S. J. van Deventer, S. W. Mallicoat, W. M. Kriven, "Microstructural Characterization of Metakaolin-based Geopolymers," *Ceramic Transactions*, vol. **165**. Advances in Ceramic Matrix Composites **X**, edited by J. P. Singh, N. P. Bansal and W. M. Kriven 71-85 (2005).

[25] P. Duxson, J. L. Provis, G. C. Lukey, S. W. Mallicoat, W. M. Kriven and J. S. J. van Deventer, "Understanding the Relationship between Geopolymer Composition, Microstructure and Mechanical Properties," *Colloids and Surfaces A – Physicochemical and Engineering Aspects*, **269** [1-3] 47-58 (2005).

[26] J. L. Bell, M. Gordon and W. M. Kriven, "Use of Geopolymeric Cements as a Refractory Adhesive for Metal and Ceramic Joins," *Cer. Eng. Sci. Proc.* Edited by D.-M. Zhu, K. Plucknett and W. M. Kriven, vol **26**, [3] 407-413 (2005).

[27] M. Gordon, J. Bell and W. M. Kriven, "Thermal Conversion and Microstructural Evaluation of Geopolymers or "Alkali Bonded Ceramics" (ABCs)," Ceramic Transactions, vol. **175**. Advances in Ceramic Matrix Composites **XI**. Edited by N. P. Bansal, J. P. Singh and W. M. Kriven, 225-236 (2005).

[28] W. M. Kriven, J. Bell, M. Gordon and Gianguo Wen, "Microstructure of Geopolymers and Geopolymer-based Materials," pp 179-183 in <u>Geopolymer, Green Chemistry and Sustainable Development Solutions</u>, edited by Joseph Davidovits. Proc. World Congress Geopolymer, 2005, St. Quentin, France. Published by the Geopolymer Institute, St. Quentin, France (2005).

[29] P. Duxson, S. W. Mallicoat G. C. Lukey, W. M. Kriven and J. S. J. van Deventer, "Effect of Alkali and Si/Al Ratio on the Development of Mechanical Properties of Metakaolin-based Geopolymers," *Colloids and Surfaces A-Physicochemical and Engineering Aspects*, **292** 8-20 (2007).

[30] Z. Rozenberg and Y. Yeshurun, "The Relation Between Ballistic Efficiency and Compressive Strength of Ceramic Tiles," *Int. J. Impact Engng.*, 7 357 (1988).

[31] F. Huang and L. Zhang, "DOP Test Evaluation on the Ballistic Performance of Armor Ceramics against Long Rod Penetration," *AIP Conference Proceedings* **845** 1383-1386 (2006).

[32] W. M. Kriven, S. J. Lee, M. A. Gülgün, M. H. Nguyen, and D. K. Kim, "Synthesis of Oxide Powders via Polymeric Steric Entrapment," (invited review paper), in *Innovative Processing/Synthesis:Ceramics, Glasses, Composites III, Cer. Trans.*, **108** 99 – 110 (2000).

[33] M. A. Gülgün, W. M. Kriven, and M. H. Nguyen, "Processes for Preparing Mixed-Oxide Powders," U. S. Pat. No. 6482387, November 19, 2002.

[34] G. R. Antis, P. Chantikul, B. R. Lawn, D. B. Marshall, "A Critical Evaluation of Indentation Techniques for Measuring Fracture Toughness : I, Direct Crack Measurements," *J. Am. Ceram. Soc.*, **64** [9] 533-38 (1981).

SOME PRACTICAL REQUIREMENTS FOR ALUMINA ARMOR SYSTEMS

K.Sujirote; K.Dateraksa; and N.Chollacoop
National Metal and Materials Technology Center
114 Thailand Science Park, Paholyotin Rd.,
Klong Luang, Patumtani 12120 THAILAND

ABSTRACT

Typical modern lightweight armor would now comprise a ceramic face layer with a more flexible backing layer. The role of the ceramic outer layer is to blunt the projectile and dissipate the load over a wide area. The dynamic load of projectile impact on the ceramic tiles involves complex mechanisms of penetration and perforation, thus making it necessary to introduce many simplifying assumptions to the problem. An experimental investigation has been conducted to investigate the ballistic performance of alumina tiles as a function of weight limitation, threat level, backing material, and multi-hit performance. The samples were impacted by 9 mm or SS109 ammunitions at impact speed of 420 and 920 m/s, respectively. Measurements of the trauma sizes on a standard clay box directly behind the armor samples could be explained by observations of debris patterns 'frozen' on the rear surface. Typically, more conoids occur with thicker sample and production of pulverized zone is a necessary condition for the penetration resistance. Debris from SS109 (sharp projectile with high impact speed) was finely comminuted and compacted at the locus of conoid coaxial cracks, whereas relatively larger debris was evident from blunt projectile. Likewise, the backing materials with different impedance effects may induce tensile failure across the boundaries, which significantly influence on crack propagation and debris comminution process. Impact resistance at edges and joints were sharply deteriorated by crack pattern alteration, leading to absence of locus conoid cracks. Thus it is necessary to maintain the same ballistic performance at the edge as well as the center of the tile in order for the armor to withstand multiple hits.

1. INTRODUCTION

It has been established in the last three decades that ceramic facing structures represent a group of light armor[1], which is significantly efficient in reducing the penetrability of a conical striker[2]. Alumina is one of the most readily available and cost-effective materials with very attractive physical properties. Material properties, such as dynamic compressive failure energy, abrasive friction and flow properties of the finely fragmented material, govern the penetration resistance of ceramics. The mechanical properties of ceramics have been continuously improved by developing better processing methods[3] and achieving desired microstructures.

Compared to monolithic armor, dual hardness armor (DHA) designs have shown improved efficiency over monolithic armors. The ceramic material receives the initial impact of the projectile and its function is to destroy the head of the projectile progressively as it tries to penetrate the composite materials. In this initial, stage a major part of the impact energy is dissipated. Then in the second stage, the base layer is made of ductile material and its main function is to absorb the residual impact energy caused by the fragmented parts of the projectile as it comes to a complete halt, thus resulting in plastic deformation of the ductile material.[4] The weight of the armor is then reduced in comparison with an armor made of steel only. This combination is especially appealing when dealing with small-to-medium armor piercing calibers.[5]

Further enhancement of properties is possible with the use of materials in more novel geometries and configurations.[6-8] The optimal design and implementation of energy-absorbing materials and structural systems require a fundamental understanding of dynamic events and failure mechanisms under extreme conditions.[9] This understanding will have a major impact on the ability of designers to tailor their protective systems for reliable performance.

Wilkins[10], Shockey[11] and Sherman [12,13] have provided notable description of the ballistic failure processes in ceramic-faced armors. According to Wilkins[10], the projectile tip is first destroyed. Simultaneously, a fracture conoid initiates at the interface between the projectile and the target. The cones that are formed spread the load of the projectile onto a relatively wide area, enabling the energy of the impact to be dissipated by the plastic deformation of a ductile backing material. The backing plate yields at the ceramic interface. The tension that results in the ceramic as it follows the motion of the backup plate initiates an axial crack. According to Schockey et al[11] and Sherman et al[12,13], first a network of radial cracks forms at the bottom surface of the tile. This is followed by growth of shear dominated cone cracks initiated at edge of contact area on the top surface. Spall cracks occur parallel to the impact surface. Analysis of penetrated and partially penetrated ceramic materials has led to the observation of a comminuted zone, also referred to as Mescall zone, near the leading edge of the penetrator. Both the resistance to comminution and the ability of the penetrator to move through the resulting comminuted ceramic particles have been identified as significant factors governing the ballistic performance of a ceramic material.[14]

Microscopy provides insight into the micro-mechanisms of failure[15] and helps the development of constitutive models. Curran et al.[16] present a micromechanical model for comminution and granular flow of ceramics under impact. Corte's et al.[17] numerically modelled the impact of ceramic-composite armor, in which they present a constitutive model for finely pulverized ceramic taking into account internal friction and volumetric expansion.

In designing armor, Florence[18] developed a model for estimating the ballistic limit, assuming that ceramic only distributes the impact load over a large area onto the backing plate, which later absorbs all the energy of impact. This classical model is often used as a general guideline for designing two-component armors. Optimization for the ballistic limit of two-component armor, under both total thickness and areal density constraints or either constraint is available in literature.[19]

Nonetheless, it appears that the theoretical model tends to be conservative, the firing results showing higher ballistic limits than the predicted values. Several practical requirements undoubtedly affect the ballistic performance. For instance, the aerial density of the armor is not constant under the total thickness criterion.[2] The relative strength of the bullets and armors is a major aspect in the ballistic simulations.[20] Also, adhesive layer showed an optimum thickness for the best performance of the lightweight protection considered.[21]

The aim of this study is to focus on practical issues from processing, assembling and weight optimization for certain threat level points of view. The scope is limited to determine how the ceramic tile with chamfered edge affect the randomly impacting projectile, at various thickness and bullet velocity. In addition, the effect of backing plate; i.e., either metal or metal with polymer composite, is investigated. The results are compared with those from numerical simulation.

2. EXPERIMENTAL PROCEDURE

Alumina hexagons with nominal area of 800 mm^2, ranging from 2.5 to 10 mm in thickness, were assembled on metal mesh placed on top of an aramid fiber layer using a synthetic rubber based cement. The assemblies, approximately 20x20 cm^2 in size wrapped by a PE layer, were then backed up by either a 1.5 mm thick steel sheet or a 5 mm thick aramid-PU composite, making ceramic-metal- and ceramic-metal-polymer layer composite samples. The annotation for samples of various thicknesses with corresponding ammunition, and bullet speed used is shown in Table 1. Generally, the type and number of layers of the backing material, including the type of adhesive as well as the bonding techniques used, strongly affected the ballistic performance. Therefore, the same bonding process used for normal production was utilized for the test samples.

Table 1. Annotation for various samples being considered

Annotation	Al$_2$O$_3$ (mm)	Stainless steel (mm)	Polymer composite (mm)	Ammunition	Areal density (g/cm^2)	No. of samples
4-A	4	1.5	-	3(A) 9 mm	2.85	2
6-A	6	1.5	-	3(A) 9 mm	3.50	2
8-A	8	1.5	-	3(A) 9 mm	3.97	2
6-B	6	1.5	-	SS109	3.50	2
8-B	8	1.5	-	SS109	3.97	3
8-C	8	1.5	5	SS109	3.97	1

Ballistic performance was tested in accordance with the NIJ 0101.30 and NIJ 0101.04 standards level 3A and 3, which specify 9 mm ammunition at the velocity of 420-430 m/s and SS109 ammunition at 920-940 m/s, respectively. The bullet velocity was measured by a chronograph. The trauma damage was measured on a Roma Platilina modelling clay attached to the back of the armor system after shooting. The damage zone of the ceramics, including ceramic fragmentation and the bullet were observed on the clay deformation.

The Fastcam-APX RS 250 KC image acquisition system was used to study the ultra high-speed phenomenon of ballistic penetration. Capturing rate at 20,000 fps and shutter speed of 1/20000 s were programmed to record a sequence of separate images at prescribed time intervals. Images were acquired from a side view to the projectile path. The high-speed camera was housed behind transparent polycarbonate shields to prevent damage due to flying debris. The damage zone of the ceramics, including ceramic fragmentation, and the bullets were observed by visual inspection and measurement.

All finite element simulations in the present study were performed using ABAQUS/Explicit solver[22]. Due to the high speed, non-linear transient responses in the solutions, an axisymmetric model was chosen for simulations of a spear-shape bullet penetrating layers of alumina and steel plates, as shown in Figure 2. Details and parameters used can be seen from earlier work[20]. In this modelling, the bullet was assumed to be a single piece with a sharp tip and a total length of 18 mm, estimating the commercial bullets in terms of diameter and mass. Meanwhile, the armor plates were assumed to be homogeneous in each layer spanning the total length of 30mm in diameter to minimize the effect from the far field condition. Complication from the finite alumina tiles was not yet taken into account for the present investigation.

3. RESULTS AND DISCUSSION

3.1 3A (9 mm) Projectile Results

Figure 3 shows detailed impact phenomenon captured by the high-speed camera during the impact of 9 mm bullet at velocities ranging from 419 to 431 m/s, as measured by the sensors after the bullets leave from the barrel (Figure 1, left). Clear photographs of the early stages of the impact phenomenon were obtained. Immediately after impact (within 50 ns), stress wave propagated forward and backward could be observed on the PE wrapping sheath. The high amplitude stress is generated as the result of the relatively large acoustic impedance of the ceramic and its much higher elastic compressive limit compared to that of the hard steel. Then ceramic ejecta were observed emerging from the front surface. The ejecta flow was radially disperse. A cone of pulverized ceramic plume is formed, surrounding the projectile. It could be seen from the movie clips that not only the impact energy and damage extent on the surface, but also the amount and velocity of the debris exploded from the surface were much more severe on thicker samples, leading to higher energy absorption.

Observation of the remaining of each specimen showed some localized damages on a single tile for the center hit or a few peripheral tiles for the edge hit, while the back steel plates show large protuberances (Table 2). Trauma depths measured from the imprints on the clay box, as shown in Fig.4, increased proportionally to the alumina thickness. The sample was considered thoroughly penetrated when the deformation was more than approximately 25 mm in depth (Fig. 4). In this case, a petal damage on the back steel plate was formed. Horsfall & Buckley[23] reported that the effect of the full width through-thickness cracks was to lower the ballistic limit velocity by only 3% compared to that of the undamaged panes. This means that the presence of cracks in a ceramic armor tile should not be sufficient reason to require replacement of the panel, a fact of some importance given the likelihood of damage in the military environment. It was proposed that the small value of the reduction in performance was due to the presence of a well bonded composite backing and a frontal spall shield. The presence of a large crack at the impact point has little effect since the ceramic in this area is extensively comminuted ahead of the projectile upon impact, enhancing the bondage with the composite backing. The backing and spall shield conserve the structural integrity of the panel, which altogether contain the radial stresses generated by the impact event.

Table 2. Plate damage after impacts

Sample annotation	Ceramic		Metal	Clay imprint
	Front	Back	Back	
4-A				
6-A				
8-A				

Debris from 4-A, 6-A, and 8-A samples frozen on the wrap-up sheath is shown in Fig.5. The tip of the projectile is eroded significantly, leaving molten lead from the ammunition on the penetrated and partially penetrated samples. According to the tile thickness and thus the energy absorbed, different types of cracks could be observed[24]. It could be seen that the locus of conoid coaxial cracks starts at the impact point; radial tensile cracks are initiated at the back surface close to the axis of the impact. Star cracks are formed at the side of conoids. Tangential spall cracks occurred due to shear stress waves reflected from the edges of the tile and due to the formation of the cone cracks; lateral spall cracks may also form due to longitudinal stress waves reflected from the backing support. These tensile- and shear-stress-derived cracks occur at relatively low stress, and it is expected to be increasingly important as the armor thickness is reduced (4 mm). On the other hand, compressive fracture takes place at fairly high stress, and it will be important when the impacted body has a large thickness.

For ceramics with high energy transmission (8 mm), a set of concentric circumferential fractures also develops. The fracture of ceramics leads to the formation of a fine powder, which deforms non-reversibly by the relative motion of ceramic granules. Thus, the friction effect becomes important and enables a comminuted ceramic support deviatoric stress. Although the energy consumed in the fracture of ceramic tiles is very limited compared with the impact energy, the development of a fracture zone ahead of the penetrator seems to be most important in defeating the projectile[2].

The energy of the systems, namely the internal energy (IE), the kinetic energy (KE) and total energy (TE), during impacts together with the corresponding contours of von Mises stress on deformed meshes at various time instants t during the impact are plotted in Figure 6. The contour plots exhibit high stresses during the first period after the initial impact at the center of the ceramic plates. If the bullets do not penetrate, as in cases 6-A and 8-A, the armors can respectively absorb 98.5, 97.4 and 97.8% of the bullet energy. Meanwhile, only 74.0% of the impact energy is absorbed in the case 4-A, in which the bullet perforates. Vanichayangkuranont et al[20] emphasised that heightened tensile strengths of ceramics with elastic bullets is necessary for the results to qualitatively agree with the experimental observations.

3.2 SS109 Projectile Results

The size and weight of SS109 ammunition is not much different from those of 3A(9mm), but it is sharper and its velocity is twice as high. The velocity (V2) and depth (U2) of the bullet tip are also shown in Fig. 7. As the velocity of the projectile is increased, both the projectile and the target break into a larger number of finer pieces, leading to more efficient energy absorption. Considerable amount of the energy transferred to the target is converted to the kinetic energy of the ejecta. The ejection process clears the pulverized ceramic away to accommodate the penetration of the projectile. A pulverized zone is formed ahead of the projectile due to intense stress conditions. Crack patterns on sample 8-B are much less fragmented than those on sample 8-A due to a smaller projectile head for SS109. Images of the plate damage on two 8-B samples after impact are shown in Table 3. Both specimens showed some localized damages on a single tile while the back steel plates showed full penetration by petal damage in sample 8-B-I and large bulge on sample 8-B-II, due to the differences in the locations of the alumina tiles impact.

The $180°$ sweep of the axisymmetric deformed plot at $t = 0.2$ ms with the steel plate profile at various time are shown in Fig. 8 for comparison with the experimental observation in Table 3. The bulge area with diameter of 50 mm predicted by the simulation in Fig. 8 is in a good agreement with the value of 50-55 mm observed in the post-firing samples in Table 3. It should be noted that this numerical simulation agrees well with the event on sample hit at the edge, but not the one hit right at the center of the ceramic tile.

Table 3. Plate damage on two 8-B samples after impacts

case	ceramic		metal	
	front	back	back	clay
8-B-I				
8-B-II				

The computational model (Fig.9) shows that alumina layer absorbs about two-thirds of the impact energy from the bullets with the remaining one-third dissipating into frictional work from the contacting interaction. The most severe damage to the armor occurs at the backside of the alumina layer due to the superposition of reflecting stress waves. From strain and total energy plots, it can be seen that the armor plate can effectively dissipate the impact energy of the bullet into the internal strain energy in each layer and the frictional work due to the contacting interactions. Note that the friction during the bullet impact is significant, representing approximately half of the internal strain energy absorbed in the participating armor layers.

It appears that the theoretical model tends to be conservative, the firing results showing higher ballistic limits than the predicted values. Referring to Fig.4, one out of three samples with 8 mm Al2O3 was not defeated by the SS109 ammunition. It is interesting to see, from a closer look at their frozen debris on the back polymer sheath, that the survived sample was hit right at the center of the tile (Fig.10a), while the defeated ones at the edge (Fig.10b). Fig. 11 shows SEM images of the fractured pieces of the two samples. It is observed that in the case of SS109, the fracture is mostly deviated along grain boundary and fragmented spalling grains are evident on the surface, which can therefore offer greater resistance to penetration. However, in the case of 3A (9 mm), the fracture surface is less defined along the paths of least resistance.

Several practical requirements undoubtedly affect the ballistic performance. First of all, due to processing point of view, the alumina tiles used in this study were chamfered, leaving the edge 1 mm thinner than the center. But it is shown that it may not be wise to barter the processing with the performance.

Secondly, the adhesive thickness[21] affects the armor performance in three aspects; i.e. shear stress, ceramic spalling, and energy absorption. The shear stress on the adhesive decreases with a thick layer, avoiding its failure and holding the ceramic material attached to the backing plate after impact. Ceramic spalling is reduced with thin adhesive layers that prevent bending of the hard tile and energy absorption by the backing plate is greater with a thick layer, which facilitates load transfer from ceramic to metal. Therefore a variation of the thickness of the adhesive layer affects the efficiency of the armor.

Thirdly, the ballistic efficiency of ceramic tiles can be vastly improved by judiciously restraining them with membranes of polymeric composites or metal sheets. Steel has a high elastic modulus and high strain energy dissipation capability. It is therefore able to reduce the deformation at impact, to dissipate the kinetic energy, and to minimise reflection of the stress waves from the tile/support interface.[8] On the other hand, polymeric composite possesses considerably lower density, giving rise to better efficiency. Cunniff[26] suggested a series of confinement states of ceramics, having different impact resistance and armor-piercing resistance, are ranked as follows:
- a free ceramic tile has low impact resistance
- a ceramic tile glued on a tough material has high resistance, which is the configuration used in this study.
- a ceramic tile under biaxial confinement has higher impact resistance;
- triaxial prestressed ceramic composite has the highest impact resistance.

4. CONCLUSIONS

The armor system based on alumina ceramic's bonded with appropriate backing materials can provide ballistic protection dependant on the ceramics and backing material thickness. The thickness of ceramics may also affect crack formation and development. Usually more conoids occur with greater thickness. Fragments of damaged ceramics with different sizes ranging from big chunks to a fine powder were observed after fracturing. Formation of very fine fragments is related to the crushing ahead of the projectile, and the coarser fragments are related to the stress wave interaction.

Numerical simulation could exploit empirical data to predict the ballistic events and to design suitable armor. The developed model agrees well with experiments at lower bullet velocity in the range of 400 m/s, level 3A. But most of the models are rather conservative, especially at high threat level; i.e. bullet velocity more than 800 m/s, which various dynamic factors is difficult to or can not be determined. In this study, practical requirements, which have not been included in the model, such as edge design, adhesive layer, and residual stress due to physical confinement have been shown to be critical.

ACKNOWLEDGEMENTS

This research is supported by the National Metal and Materials Technology Center (MTEC), contract no. MT-B-48-CER-07-188-I. Special thanks are due to Dr. K. Maneeratana, Dr. K. Prapakorn, Assc. Prof. Dr. T. Amornsakchai, Asst. Prof. Dr. S. Rimdusit and Maj. Gen. Dr. W. Phlawadana.

REFERENCES

1. Rapacki,E.J.; Hauver,G.E.; Netherwood,P.H.; Benck,R.F. 'Ceramics for armors - a material system perspective' 7th Annual TARDEC Ground vehicle survivability symposium, USA, 1996
2. Wang,B. & Lu,G.' On the optimisation of two-component plates against ballistic impact' Journal of Materials Processing Technology 57 (1996) 141 145
3. Forquin,P.; Tran,L.; Louvigne,R.F.; Rota,L.; Hild,F. 'Effect of aluminum reinforcement on the dynamic fragmentation of SiC ceramics' Int.J.Impact Eng. 2003, 28, 1061-76
4. Goncalves,D.P.; de Melo,F.C.L.; Klein,A.N.; & Al-Qureshi,H.A. 'Analysis and investigation of ballistic impact on ceramic/metal composite armor' Int.J.Impact Eng. 2004, 44, 307-16
5. Hauver,G.E.; Netherwood P.H.; Benck,R.F.; Kecskes,L.J. 'Enhanced ballistic performance of ceramic targets' 19th Army Science Conference, USA 1994
6. Shermon D. 'Impact failure mechanisms in alumina tiles on finite thickness support and effect of confinement' Int.J.Impact Eng. 2000, 24, 313
7. Vaidya,U.K.; Abraham,A.; & Bhide,S. 'Affordable processing of thick section and integral multifunctional composites' Composites: Part A 32(2001) 1133-42
8. Sarva,S.; Nemat-nesser,S.; McGee,J.; & Isaacs,J. 'The effect of thin membrane restraint on the ballistic performance of armor grade ceramic tiles' Int.J.Impact Eng. in press
9. Shih,C.J. & Adams,M.A. 'Ceramic array armor' US 6,532,857 B1, 2003
10. Wilkins ML. Mechanics of penetration and perforation. International Journal of Engineering Science 1978;16:793–807.
11. Shockey DE, Marchand AH, Skaggs SR, Cort GE, Burkett MW, Parker R. Failure phenomenology of confined ceramic targets and impacting rods. International Journal of Impact Engineering 1990;9(3):263–75.
12. Sherman D, Brandon DG. The ballistic failure mechanisms and sequence in semi-infinite supported alumina tiles. Journal of Materials Research 1997;12(5):1335–43.
13. Sherman,D.; Ben-Shushan,T. 'Quasi-static impact damage in confined ceramic tiles' Int.J.Impact Engng 1998; 21, 245-65
14. Stepp,D.M. 'Damage mitigation in ceramics: Historical developments and future directions in army research' Ceramic Armor Materials by Design ACerS 2002, 421-28
15. Viechnicki,D.J.; Slavin,M.J.; and Kliman,M.I. 'Development and current status of armor ceramics' Am.Cer.Soc.Bull. 70[6] 1035-39(1991)

16. Curran DR, Seaman L, Cooper T, Shockey DA. Micromechanical model for comminution and granular flow of brittle material under high strain rate application to penetration of ceramic targets. Int J Impact Eng 1993;13(1):53.
17. Corte´ s R, Navarro C, Martinez MA, Rodriguez J, Sanchez-Galvez V. Numerical modelling of normal impact on ceramic composite armors. Int J Impact Eng 1992;12(4):639.
18. Florence, A.L. Interaction of projectiles and composite armor, Part 11, Standford Research Institute, Menlo Park, California, AMMRG-G-69-15, 1969.
19. Shi, J. & Grow, D. 'Effect of double constraints on the optimization of two-component armor systems' Composite Structures, in press
20. Vanichayangkuranont,T.; Maneeratana, K.; & Chollacoop,N. 'Numerical Simulations of Level 3A Ballistic Impact on Ceramic/Steel Armor' The 20th Conference of Mechanical Engineering Network of Thailand, 18-20 October 2006, Nakhon Ratchasima, Thailand
21. Lo´ pez-Puente, J.; Arias,A.; Zaera,R.; Navarro,C. 'The effect of the thickness of the adhesive layer on the ballistic limit of ceramic/metal armors. An experimental and numerical study' International Journal of Impact Engineering 32 322 (2005) 321–336
22. Habbitt, Karlsson and Sorensen Inc., 2003. ABAQUS version 6.3 Theory Manual, Pawtucket.
23. Horsfall,l. & Buckley,D. 'The effect of through-thickness cracks on the ballistic performance of ceramic armor systems' Int.J.Impact Eng. 18(3)309-18
24. Sujirote,K. and Dateraksa,K. 'Ballistic Fracture of Alumina Ceramics' MSaT IV, Thailand
25. Anderson,Jr.,C.E. & Walker,J.D. 'An analytical model for dwell and interface defeat' Int J Impact Eng 31 (2005)1119-32
26. Cunniff PM 'Assessment of small arms (ball round) body armor performance' Proceedings 18th Int.Symp. On Ballistics 1999, p.806

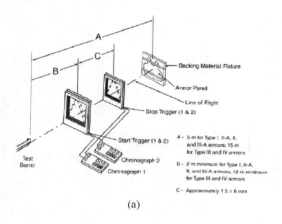

(a)

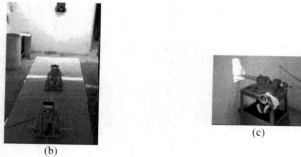

(b)

(c)

Fig.1 Ballistic experimental set up (a) for NIJ standard 0101.04, (b) for actual chronograph and (c) gun barrel used in this study.

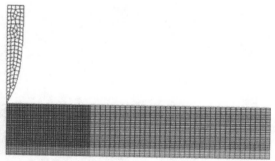

Fig 2. Mesh design for the bullet and central layers of alumina and steel (red).

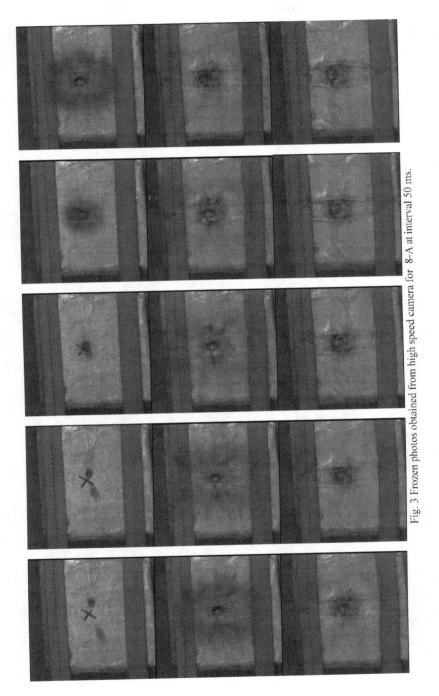

Fig. 3 Frozen photos obtained from high speed camera for 8-A at interval 50 ms.

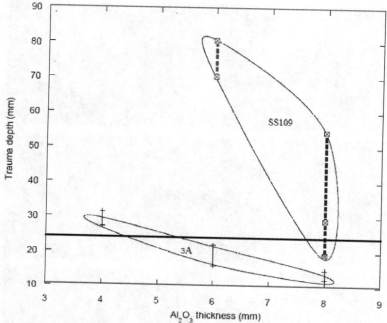

Fig. 4 Graph exhibiting trauma depth VS thickness for 3A and SS109 ammunition.

Fig. 5 Frozen debris for 4-A, 6-A, and 8-A, respectively

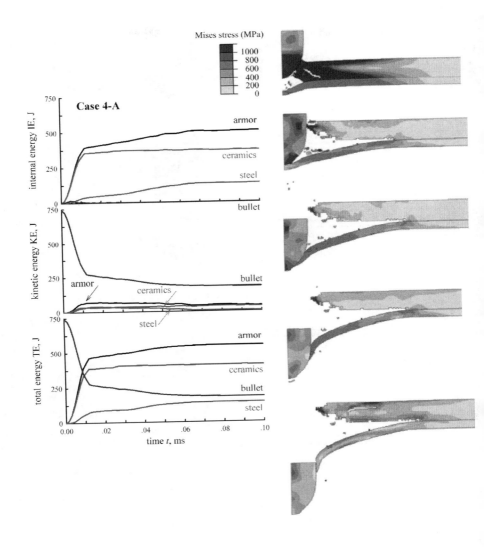

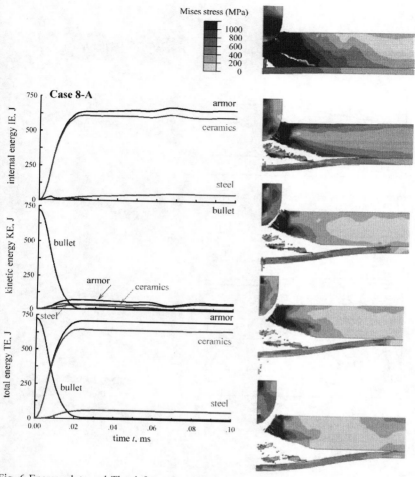

Fig. 6 Energy plots and The deformation and von Mises stress at time instant t = 0.01, 0.03, 0.05, 0.07, and 0.09 ms. For 4-A and 8-A

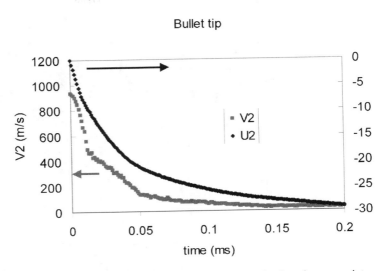

Figure 7. The velocity and depth of bullet penetrating into the armor plate

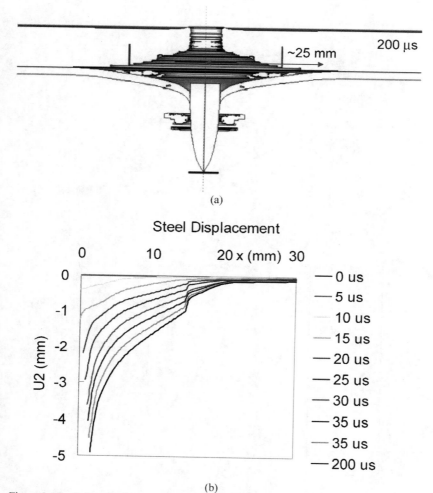

(b)

Figure 8. The deformed plot at t = 0.2 ms for bulge area measurement (a) with the steel layer profile at various time (b)

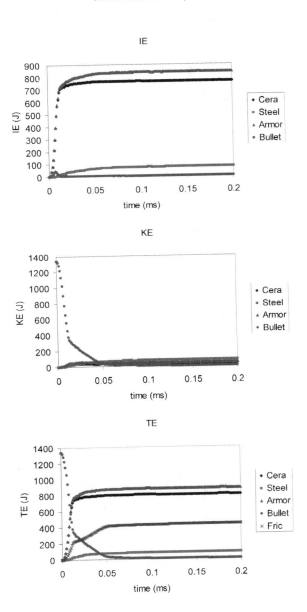

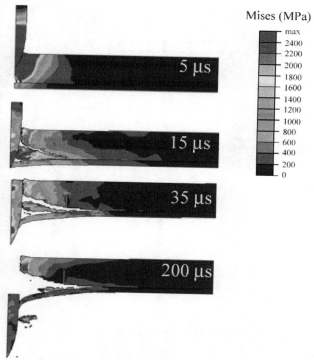

Mises (MPa)

Fig 9. Energy plots and the deformation and von Mises stress at time instant t = 0.005, 0.015, 0.035 and 0.2 ms for the case 8-B

(a)   (b)   (c)

Fig. 10 Frozen debris for 8-B (a) center, (b) edge, and (c) for 8-C.

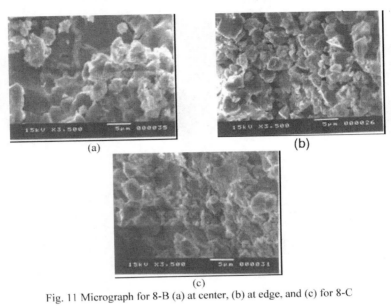

(a)

(b)

(c)

Fig. 11 Micrograph for 8-B (a) at center, (b) at edge, and (c) for 8-C

Glasses and
Transparent Ceramics

LIMIT ANALYSES FOR SURFACE CRACK INSTANTANEOUS PROPAGATION ANGLES IN ELASTIC HERTZIAN FIELD

Xin Sun and Mohammad A. Khaleel
Pacific Northwest National Laboratory
K6-08
PO BOX 999
Richland, WA 99352

ABSTRACT

This paper presents the limit analyses of instantaneous crack propagation angle in the classical linear-elastic Hertzian field. The formation and propagation of Hertzian cone crack have been studied extensively in the open literature. Many experiments have been carried out on soda-lime glass under static loading. The observed final Hertzian cone angle is usually around 22°, and it tends to increase with increasing indentation velocity. The goal of this paper is to use the crack tip stress field derived in linear elastic fracture mechanics (LEFM) to bound the instantaneous surface ring crack extension angles in the Hertzian field. The upper and lower bounds for the instantaneous crack extension angle are expressed in terms of the ratios of the mode I and mode II stress intensity factors at the crack tip field. It is shown that under a pure mode I dominant stress field, instantaneous crack growth would follow the original surface crack direction; and under a perfect mode II stress field, the instantaneous crack extension angle has a limiting angle of 19.5°. The actual instantaneous crack extension angle will fall between these two limit states depending on the ratio of K_I/K_{II} at the crack tip.

INTRODUCTION

The problem of a hard spherical indentor being pressed on to a semi-infinite elastic field is referred to as the classical Hertzian indentation problem following the original work of Hertz over a century ago, see Figure 1 [1]. It is generally accepted that Hertzian fracture begins as a surface ring crack outside the elastic contact and then, at a critical load, propagates downward and flares outward into a stable, truncated cone configuration [2]. There is an enormous amount of literature on the Hertzian indentation system, and on the conditions for formation and propagation of Hertzian cone cracks [2-6]. Interested readers should refer to Refs. 2-5 and the articles cited therein. For example, in their seminal work in 1967, Frank and Lawn [6] applied the Griffith fracture criterion and calculated the amount of mechanical energy released in terms of the preexisting stress fields as a function of crack length for an assumed crack geometry. They stated, 'to an excellent first approximation', that 'cracking proceeds orthogonally to the greatest tensile stress, thus following a surface delineated by the trajectories of the other two principal stresses' in the preexisting field. Frank and Lawn compared the angles of the cone crack and of the preexisting stress trajectory, and found that they were essentially identical: about 22° from the surface of the glass. However, for reasons of expediency, they used a value of 0.33 as the Poisson's ratio for glass in stress calculations. This Poisson's ratio is much higher than that of the soda–lime glass studied, which is typically around 0.21. A value of 0.21 would lead to a Hertzian cone angle prediction of greater than 30°, which is in poor agreement with established experimental observations.

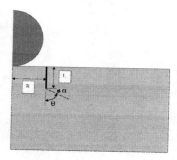

Figure 1. Schematic illustration of the Hertzian indentation system

Later, Lawn, Wilshaw, and Hartley [7] used a computer model to simulate the growth rate of cone cracks, by applying fracture mechanics criteria to advance the cracks incrementally. In that work, it was also assumed that the crack followed those trajectories defined by the preexisting stress field. Those researchers also noted the discrepancy between the angle of the preexisting stress trajectories (>30°) and the observed crack angle of 22° for soda–lime glass for which the literature values of Poisson's ratio were between 0.20 and 0.25. They noted that it would be necessary to assume a Poisson's ratio of 0.33 to get satisfactory agreement between the experimental crack angle and the preexisting stress trajectory. When recognizing the unsatisfactory agreement between the angles, the authors questioned whether it was appropriate to calculate the stress fields on the basis of isotropic, linear elasticity theory, and speculated that the magnitude of the stress fields, as well as their anisotropic geometry might lead to an "effective" Poisson's ratio, locally at the crack tip, which is very different from the value far from the crack tip.

In fact, the stress field underneath the indentor strongly depends on the Poisson's ratio of the elastic field [8]. For example, Warren [9] studied the shape of the stress trajectory as a function of Poisson's ratio. For a value of 0.33, he also obtained the angle of 22° for trajectories as had Lawn et al [7]. For a Poisson ratio of 0.19, the angle of the trajectory was found to be greater than 30°. Therefore, the concept was reinforced that the quantitative predictions of fracture behavior are very dependent on the value chosen for Poisson's ratio.

Since the stress field evolves with the propagation of the cone crack, question was then raised about whether cone crack should, indeed, follow the preexisting stress trajectory [10]. Of course, many researchers had recognized that the presence of the crack changed the stresses in the solid. For example, Frank and Lawn [6] referred to the "extremely different stresses present" in the solid when a crack exists, and stated that 'we must therefore deny the hypothesis that it (crack) follows stress trajectories exactly'. They also stated that when crack 'enters a region in which it is obliquely stressed', it will 'suffer deflexion in the direction readily inferred from Griffith's case'.

In order to examine the effects of pre-existing surface cracks on Hertzian cone propagation angles, Kocer and Collins [11] used finite element method to model the surface crack growth in the Hertzian stress fields. In their work, the crack is incrementally advanced along the direction of maximum strain energy release, as calculated by the evolving, rather than the preexisting stress fields. For the modeled Hertzian indentation system, a cone crack is observed to grow, but at an angle which is significantly different from that defined by the normal to the maximum preexisting tensile stress. With very fine mesh at the crack tip and the realistic Poisson's ratio of 0.21 for soda-lime glass, the predicted angle of the cone crack propagation is in excellent agreement with observations on experimentally grown cone cracks in glass.

In this study, we take into consideration the existence of a ring crack on the surface of a semi-infinite glass field and study the limits for ring crack instantaneous propagation angles using LEFM crack tip stress field solutions under Hertzian indentation load. The stress field formulation is based on Erdogan and Sih's [12] work over 40 years ago on crack extension in plates. It is shown that under mixed mode loading conditions, assuming that crack extension starts at its tip in radial direction and that it follows the plane perpendicular to the direction of greatest tension at the crack tip, the ratio of K_I/K_{II} at the tip of the surface ring crack determines its instantaneous propagation angle. Under pure tensile stress field, that is, when the ratio of K_I/K_{II} is very high at the crack tip, ring crack first extends vertically downward. On the other hand, if $K_I/K_{II} \sim 0$, i.e., when crack tip is dominated by K_{II}, ring crack first extends at a limiting angle of 19.5°. For any mixed fields as in the actual Hertzian indentation field, the instantaneous propagation angle would fall between 19.5° and 90°. In other words, this study provides the theoretical bound analyses on the first increment of the crack extension angle for the numerical solutions provided by Kocer and Collins [11] under different loading conditions. Several numerical examples are then presented illustrating the effects of the surface crack location on its instantaneous propagation angle using the above proposed propagation criterion.

LINEAR ELASTIC CRACK TIP SOLUTIONS IN HERTZIAN FIELDS

Under plane strain or generalized plane stress where material contains a straight crack, the stress field in the neighborhood of the crack tip can be expressed as follows (see Figure 2) following Erdogan and Sih [12]:

$$\sigma_r = \frac{1}{(2\pi r)^{1/2}} \cos\frac{\theta}{2} [K_I(1+\sin^2\frac{\theta}{2}) + \frac{3}{2}K_{II}\sin\theta - 2K_{II}\tan\frac{\theta}{2}] \tag{1}$$

$$\sigma_\theta = \frac{1}{(2\pi r)^{1/2}} \cos\frac{\theta}{2} [K_I \cos^2\frac{\theta}{2} - \frac{3}{2}K_{II}\sin\theta] \tag{2}$$

$$\tau_{r\theta} = \frac{1}{2(2\pi r)^{1/2}} \cos\frac{\theta}{2} [K_I \sin\theta + K_{II}(3\cos\theta-1)] \tag{3}$$

where K_I and K_{II} are the mode I and mode II stress intensity factors at the crack tip.

Assuming the pre-existence of a ring crack on the surface of a semi-infinite glass field, the above crack tip stress field solutions can be adapted to the more commonly used Hertzian cone crack system, as shown in Figure 3, through coordinate transformation:

$$\sigma_r = \frac{1}{(2\pi r)^{1/2}} \cos\frac{\theta}{2} [K_I(1+\sin^2\frac{\theta}{2}) - \frac{3}{2}K_{II}\sin\theta + 2K_{II}\tan\frac{\theta}{2}] \tag{4}$$

$$\sigma_\theta = \frac{1}{(2\pi r)^{1/2}} \cos\frac{\theta}{2} [K_I \cos^2\frac{\theta}{2} + \frac{3}{2}K_{II}\sin\theta] \tag{5}$$

$$\tau_{r\theta} = \frac{1}{2(2\pi r)^{1/2}} \cos\frac{\theta}{2} [K_I \sin\theta - K_{II}(3\cos\theta-1)] \tag{6}$$

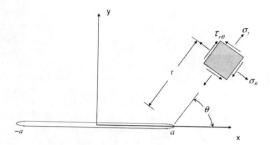

Figure 2. Stress components near the crack tip in the cylindrical coordinate system

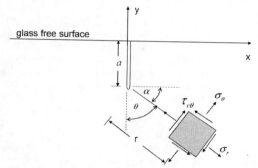

Figure 3. Hertzian cone in axisymmetric coordinate system

This is done by utilizing the axisymmetric nature of the Hertzian indentation field and its equivalence to the case of generalized plane loading as discussed by Erdogan and Sih [12]. Next, two commonly recognized hypotheses for the extension of cracks in a brittle material under slowly applied plane loading are used [12]:

- The crack extension starts at its tip in the radial direction;
- The crack extension starts in the plane perpendicular to the direction of greatest tension.

These hypotheses were used by Erdogan and Sih [12] in studying instantaneous crack extension in plates under mixed mode loading, and they imply that the crack will start to grow from the tip in the direction along which the crack tip local tangential stress σ_θ is maximum and the shear stress $\tau_{r\theta}$ is zero. It then follows that the angle of maximum tangential stress ($\frac{d\sigma_\theta}{d\theta} = 0$) and zero shear stress can be found from:

$$\cos\frac{\theta}{2}[K_I \sin\theta - K_{II}(3\cos\theta - 1)] = 0$$

which gives
$$\theta_0 = \pm\pi, \ or$$
$$K_I \sin\theta_0 - K_{II}(3\cos\theta_0 - 1) = 0$$

The non-trivial solution is then by finding the θ_0 satisfying

$$\frac{K_I}{K_{II}} = \frac{3\cos\theta_0 - 1}{\sin\theta_0} \tag{7}$$

Figure 4 depicts the relationship between K_I / K_{II} and θ_0 for $0° < \theta_0 < 180°$ governed by equation (7). The conventional Hertzian cone extension angle α_0 (see Figure 2) is then obtained as

$$\alpha_0 = 90° - \theta_0 \tag{8}$$

If the crack tip field is mode I dominant, i.e., $K_I / K_{II} >> 0$, then $\theta_0 = 0$ which indicates that the instantaneous crack propagation would follow the direction of the pre-existing surface ring crack. On the other hand, if the crack tip field is mode II dominant, i.e., $K_I / K_{II} = 0$, then $\theta_0 = 70.5°$ which leads to $\alpha_0 = 19.5°$. Therefore, the actual instantaneous crack extension angle depends on the crack tip load mode mixity.

Since the strain energy release rate and stress intensity factors at the crack tip are directly related in LEFM, this study can be considered as the theoretical bound analyses for the numerical study presented by Kocer and Collins [11].

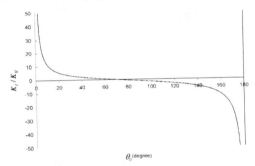

Figure 4. K_I / K_{II} versus θ_0 described by equation (7)

SURFACE CRACK PROPAGATION IN ELASTIC HERTZIAN FIELD

In this section, we use finite element analyses to illustrate different instantaneous crack propagation angles as a function of K_I/K_{II}, as described in the previous section, and to compare with the numerical crack propagation results presented by Kocer and Collins [11]. Figure 5 shows the finite element mesh used in computing the classical Hertzian stress field in which a semi-infinite glass block is being indented by a 1mm-diameter rigid sphere. In this model, no pre-existing surface crack is assumed in the classical Hertzian field. As discussed earlier, such a linear elastic stress field under a rigid spherical indentor has been studied by many researchers [1-6]. Figures 6 shows the predicted contour of maximum principal stress at indentation depth of 0.01mm using Poisson's ratio of 0.21. The predicted maximum principal stress field is very similar to those analytical solutions described by Frank and Lawn [6] as well as Mouginot and Maugis [8]: a very small, concentrated zone right underneath the contact region is predicted with three-dimensional compressive state of stress. The highest maximum tensile stress exists on the surface of the glass block, immediately outside the contact radius. This high tensile stress has been attributed to the formation of ring-cracks from pre-existing surface flaws [2-6].

In addition to the principal stress field, it is also interesting to note that a concentrated region with very high magnitude shear stress is developed inside the glass block, underneath the contact radius, see Figure 7. Moreover, the shear stress values in this region are orders of magnitude higher than the corresponding tensile stress.

According to equations (7)-(8), if the tip of the surface crack resides in this high shear zone, its instantaneous propagation angle will be more driven by mode II. Therefore, the exact instantaneous propagation angle of a surface crack will depend on the location of the crack tip in the indentation stress field and its associated crack-tip loading mode mixity, i.e., the ratio of K_I/K_{II}.

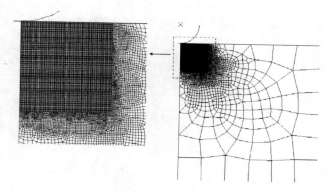

Figure 5. Axisymmetric finite element mesh used in the Hertzian stress field simulation

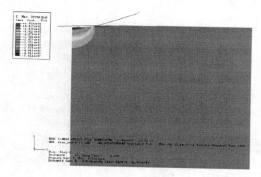

Figure 6. Contour of maximum principal stress field in classical Hertzian indentation

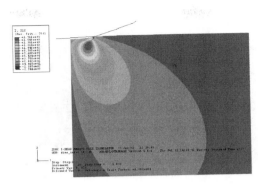

Figure 7. Contour of in-plane shear stress field in classical Hertzian indentation

Next, a series of linear elastic finite element analyses were performed in which an initial surface ring crack is introduced and its location with respect to the contact radius is systematically varied: the distance x from the ring crack to the contact outer edge increases from 0 to 0.086mm, see Figure 8. As in Kocer and Collins [11], the initial depth of the ring crack is assumed to be 5μm. A uniform contact pressure of 100MPa is applied on the glass surface within the contact radius. Figure 9 shows the detailed finite element mesh used around the crack tip for stress intensity calculations.

Figure 10 depicts the relationship between the calculated ratios of K_I/K_{II} at the crack tip and the corresponding instantaneous cone crack extension angles α (calculated from equations 7-8) with respect to x for all the cases considered. For example, at the very edge of the contact radius, the ratio of K_I/K_{II} for a 5μm deep crack is around 0.72, lowest among all the cases. And the corresponding instantaneous crack extension angle α_0 is found to be 32°. With increasing distance x between the crack and the contact edge, the crack instantaneous propagation angle α_0 increases very rapidly to the limiting value of 90°. Results in Figure 10 also indicate that surface cracks closer to the contact edge have more tendency to deviate from its original imposed angle ($\alpha = 90°$) and flare into Hertzian cones. Further away from the contact edge, the crack tip field becomes more mode-I driven, and surface cracks will need to first propagate more into the solids in its original imposed direction ($\alpha = 90°$) before loosing the K_I dominance at the crack tip.

These conclusions are consistent with the results shown in Figures 2 and 3 of Kocer and Collins [11] in which the initial propagation angle for the original surface flaw away from the contact edge is predicted to be around 90°. In other words, the crack grows normal to the surface initially. In Kocer and Collins [11], the crack grows vertically at $\alpha = 90°$ for about 40μm before changing directions. Moreover, the surface crack closer to the contact edge flares into Hertzian cone with much shorter vertical propagation distance (Figure 3 top in Kocer and Collins [11] with no forced deviation) than the surface crack away from the contact edge (Figure 3 bottom in Kocer and Collins [11]). These results can also serve as a qualitative validation of the conclusions reached in this study.

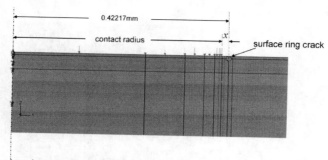

Figure 8. Schematic axisymmetric model illustrating contact radius and location of the surface crack

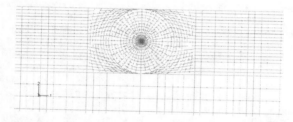

Figure 9. Detailed finite element mesh at the crack tip for stress intensity calculations

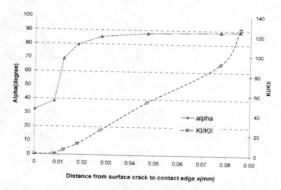

Figure 10. Relationship of K_I/K_{II} and instantaneous crack extension angle α_0 with distance x for the Hertzain field depicted in Figure 8

CONCLUSIONS

This study employs the crack tip stress field solutions in classical linear elastic fracture mechanics to bound the surface ring crack instantaneous extension angles in a static Hertzian indentation field. It is shown that the instantaneous surface crack extension angles in a Hertzian field depend strongly on the loading mode mixity around the crack tip. If the crack tip is in a mode I dominant field, instantaneous crack extension angle is 0° which means that the crack will initially propagate along the direction of the pre-existing surface crack. If the crack tip is under pure mode II field (shear dominant), the limiting angle for instantaneous crack propagation is $\theta = 70.5°$, which corresponds to the conventional notation of $\alpha = 19.5°$. Since the actual load mixity at the crack tip depends strongly on the depth of the pre-existing ring crack and the Poisson's ratio of the field, the actual angle of instantaneous crack growth would fall between $19.5° < \alpha < 90°$. The above conclusions are then illustrated with several finite element analyses of the crack tip field. Correlations with reported data in the open literature are also presented as qualitative validation of the current study.

Utilizing the fact that static indentation can be a well-controlled, stable process to quantify instantaneous crack growth angle, a coupled experimental/analytical approach to quantify fracture toughness under mixed-mode loading for brittle materials can be envisioned based on the results of this study. First, narrow surface notches of various depths can be cut into the ceramic surface using a slicing blade. The sample block should be sufficiently thick in the out-of-plane dimension to ensure a plane-strain indentation loading condition. Using a similar experimental setup as illustrated in Figure 1, the only parameters one needs to measure are the critical indentation depth, D_{crit}, or the critical contact radius, a_{crit}, at which crack extension occurs. This can be achieved with either an acoustic measuring device or by monitoring the compliance of the load versus displacement curve. D_{crit} or a_{crit} can then be used as inputs to the finite element analyses to calculate the critical values of K_I and K_{II} at fracture. Using different combinations of R and l, the fracture toughness map under various degrees of mixed mode loading can be obtained in the K_I and K_{II} space.

The analytical results presented here can also help explain the experimentally observed location-dependent surface crack propagation angles under projectile impact as shown in Figure 11. Further applications of this work should focus on engineering fracture toughness under mixed mode loading for advanced armor materials, such that surface spallation and fracture paths under projectile impact can be predicted and controlled to maximize armor protection.

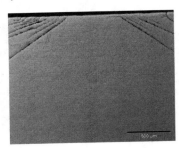

Figure 11. Micrograph of polished cross section of a recovered SiC-N cylinder subjected to spherical impact at 63m/s (replot from LaSalvia [13])

ACKNOWLEDGMENTS
Pacific Northwest National Laboratory is operated by Battelle for the U.S. Department of Energy under contract DE-AC06-76RL01830. This work was in part funded by the Dept. of Energy Office of FreedomCAR and Vehicle Technologies under the Automotive Lightweighting Materials Program managed by Dr. Joseph Carpenter.
The authors also would like to acknowledge fruitful technical discussions with Dr. Said Ahzi with University of Strasbourg, France as well as finite element calculations provided by Mr. Omar Oussouaddi with University of Metz, France.

REFERENCES
[1] H. Hertz, "On the Contact of Elastic Solids" (in Ger.) *Zeitschrift für die Reine und Angewandte methematik*, **92**, 156-171 (1881). English translation in Miscellaneous Papers (translated by D.E. Jones and G. A. Schott); pp. 146-62. Macmillan, London, UK, 1896.
[2] B. Lawn, "Indentation of Ceramics with Spheres: A Century after Hertz," *J. Am. Ceram. Soc.* 81 [8] 1977-1994 (1998).
[3] A. A. Griffith, "The Phenomena of Rupture and Flow in Solids," *Philos. Trans. R. Soc. London*, **221**, 163-198 (1920).
[4] R.F. Cook and G.M. Pharr, "Direct Observation and Analysis of Indentation Cracking in Glasses and Ceramics," *J. Am. Ceram. Soc.*, 73 [4] 787-817 (1990).
[5] B. Lawn and R. Wilshaw, "Review Indentation Fracture: Principles and Applications," *J. Mater. Sci.*, **10**, 1049-1081 (1975).
[6] F.C. Frank and B.R. Lawn, "On the Theory of Hertzian Fracture," *Proc. R. Soc. London*, A299, 291-306 (1967).
[7] B.R. Lawn, T.R. Wilshaw and N.E.W. Hartley, "A Computer Simulation Study of Hertzian Cone Crack Growth," *Int. J. Fracture*, **10** [1] 1-16 (1974).
[8] R. Mouginot and D. Maugis, "Fracture Indentation Beneath Flat and Spherical Punches," *J. Mater. Sci.*, **20**, 4354-4376 (1985).
[9] R. Warren, "Measurement of the Fracture Properties of Brittle Solids by Hertzian Indentation," *Acta Metall.*, 26, 1759-1769 (1978).
[10] E.H. Yoffe, "Stress Fields of Radial Shear Tractions Applied to an Elastic Half-Space," *Philos. Mag. A*, 54, 115-129 (1986).
[11] C. Kocer and R.E. Collins, "Angle of Hertzian Cone Cracks," *J. Am. Ceram. Soc.*, 81 [7] 1736-1742 (1998).
[12] F. Erdogan and G.C. Sih, "On the Crack Extension in Plates Under Plan Loading and Transverse Shear", *J. Basic Eng., Transactions of the ASME*, 519-527, Dec. 1963.
[13] J.C. LaSalvia, M.J. Normandia, H.T. Miller and D.E. MacKenzie, "Sphere Impact Induced Damage in Ceramics: I. Armor-Grade SiC and TiB2", *Advances in Ceramic Armor*, edited by J. J. Swab. *Proceedings of the 29th International Conference on Advanced Ceramics and Composites*, Cocoa Beach, Florida, (2005).

DYNAMIC FAILURE OF A BOROSILICATE GLASS UNDER COMPRESSION/SHEAR LOADING

X. Nie and W. Chen
Schools of Aero/Astro. and Materials Engineering
Purdue University
315 N. Grant St., West Lafayette, IN 47907

ABSTRACT
   Introduction of additional shear stress to a brittle material under dynamic uniaxial compression has been suspected to significantly affect the failure behavior of the material deforming at high rates. To examine the effects of shear stress on the dynamic failure behavior of a borosilicate glass, dynamic compression/shear experiments at the average strain rate of 250s⁻¹ are conducted using a modified Split Hopkinson Pressure Bar (SHPB). To introduce additional shear stress, cuboid specimens with the material axis inclining to the loading direction at different angles (0°, 3°, 5° and 7°) are used in the SHPB test section. To check the effects of Poisson's ratio mismatch at the specimen/platen interfaces, platens with different Young's moduli and Poisson's ratios are used and results compared. A high-speed digital camera, synchronized with the loading stress pulse, is used to record the dynamic crack initiation and propagation. Experimental results show that the equivalent stress at failure decreases with increasing shear portion in the stress. Digital images show that the cracks initiate randomly in the right specimen, whereas cracks initiate from the stress concentrated corners in the inclined specimens. Subsequent crack propagation is along the specimen axis rather than the compressive loading direction.

INTRODUCTION

   High strength glasses are used as components in transparent armors or shields, which are subjected to high speed impact loading. During the penetration process, strain rate in the target material varies significantly depending on the relative location to the penetration site, e. g. the strain rate near the penetrator head could be as high as 10^5s^{-1}, while in the crack propagation area which is relatively far from the penetrator, the strain rate is in the magnitude of 10^2s^{-1}. The effective use of glasses in the optimized structures requires a thorough understanding and a quantitative description of the mechanical response and failure behavior of glasses at a vast range of rates of deformation under complicated stress states. To develop a better understanding of glass response and failure under impact loading conditions, recent research efforts have been invested in the determination of their constitutive and failure behaviors. In pressure/shear plate impact experiments on Soda-lime glass, it was shown the introduction of additional shear component affected the one dimensional strain wave propagation and subsequent failure[1-3]. The drastic strength degradation due to the presence of shear stress indicates that the shear stress component may be critical to the deformation and failure in glasses. Similar behavior was observed on a variety of glasses[4, 5]. When a glass fails under impact loading, the stress state in the material is typically very complicated. It is therefore desired to characterize dynamic failure of glasses under complex stress states for the development of rate-dependent material models and failure criteria. However, the role of shear component on the failure is not completely understood. Recently, modified split Hopkinson pressure bar (SHPB) experiments with lateral confinement on glass samples were performed[6] in attempt to achieve a multi-axial stress state. Compared to the typical uniaxial stress loading generated by a SHPB, the introduction of lateral confinement reduces the effective shear stress in the specimen. To study the effects of higher shear stress component on the dynamic failure of glasses, new experimental techniques need to be introduced.

In this study, we present a new technique to introduce higher shear stress component in the specimen under axial compression by SHPB. The introduction of shear is achieved by employing a prismatic specimen geometry with its axis inclined to the SHPB compression loading axis (0°, 3°, 5° and 7°). To observe the failure process in the glass specimens under combined compression/shear loading, a high-speed digital camera is used to capture the entire crack initiation and propagation processes. The strength at failure in the inclined specimens is calculated using a numerical simulation that matches the experimental measurements in the bars. To examine the effects of Poisson's ratio mismatch at the specimen/platen interfaces, different platen materials are used and results are compared. It is found that the equivalent stress at failure decreases with increasing shear stress component in the specimens. After initiation, cracks in the glass specimen propagate along the specimen axis direction. The tungsten carbide and tool steel platens used in this research all have less lateral deformation than the glass specimen and thus do not affect the crack behavior in the specimen significantly.

MATERIALS AND SPECIMENS

The borosilicate glass used in this research is provided by US Army Research Laboratory, Aberdeen Proving Ground, MD, in the form of flat plates. The physical and mechanical properties of interest are: density $\rho = 2.21$ g/cm^3, Young's modulus $E = 61$ GPa, Poisson's ratio $v = 0.19$, longitudinal wave speed $C_L = 5508$ m/s, shear wave speed $C_S = 3417$ m/s. Specimens were cut from a glass plate, and then ground to specified dimensions. The specimens were 9×9 mm in cross section and 12.5 mm in length. The specimen shape is shown in Fig. 1. Four cuboid specimen geometries, with the tilting angle of 0°, 3°, 5° and 7°, respectively, were used in the dynamic compression experiments. All the specimens were polished to minimize the influence of surface flaws on measured glass strength.

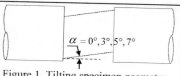

$\alpha = 0°, 3°, 5°, 7°$

Figure 1. Tilting specimen geometry (side view).

Furthermore, polished platens were placed in between the glass specimen and Hopkinson bar ends. The purposes of using these platens were to prevent the bars from being indented by the high strength glass fragments and to minimize friction on the specimen-bar interface. A mismatch in Poisson's ratio between the platens and the specimen may accelerate or restrict the axial crack initiation in the specimen from the interface under axial loading. When the lateral (radial) strain in the platen is higher than that in the specimen, the platen will deform more laterally under axial compression, which provides an additional driving force for the cracks to initiate at the end face of the specimen. If the platen has a smaller radial strain, axial cracks initiated from the end face in the specimen may not be able to propagate due to the platen restriction on crack opening. To examine the effects of this Poisson's ratio mismatch, we used platen materials of a tungsten carbide (WC) and a tool steel. To maintain the axial acoustic impedance along the loading axis of the SHPB, the diameter of the steel platens is the same as the bars, whereas the diameter of the WC platens is reduced to compensate for their higher density and higher wave speed.

EXPERIMENTAL SETUP

We modified a split Hopkinson pressure bar (SHPB), which has been commonly utilized in the high-strain-rate testing of materials to provide a complete family of dynamic stress-strain curves as a function of strain rates[7]. Originally developed by Kolsky[8], this technique was initially used for the characterization of the dynamic flow behavior of ductile materials at strain rates up to $10^4 s^{-1}$. When this device is used to determine the dynamic properties of other materials, modifications are needed to ensure that the specimen deforms under desired testing conditions. In the testing of brittle materials such like ceramics and glasses, pulse shapers are adopted to ramp the incident pulse so as to achieve a constant loading rate as well as dynamic stress equilibrium. By doing this, the maximum

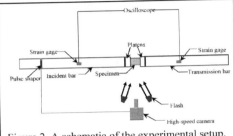

Figure 2. A schematic of the experimental setup.

achievable strain rate has to be compromised because of the linear elastic nature of those materials. In current research we used a pulse shaper to generate nearly linear ramp pulses to load the glass specimen, see Fig. 2. The annealed copper disc pulse shapers used were 9 mm in diameter and 1.7 mm in thickness to generate a linear incident ramp at the stress rate of approximately 4.2×10^6 MPa/s, which facilitates an average strain rate of $250 s^{-1}$ in the specimen. We also used disposable tool steel platens (Young's Modulus: 200GPa, Poison's ratio: 0.29) between the glass specimen and the maraging steel bar end faces to prevent damage to the bar ends by the broken glass specimen. WC platens (Young's Modulus: 566GPa, Poison's ratio: 0.20) were also used to study the Poisson's ratio mismatch effects.

To record the high-speed deformation/failure processes in the glass specimens, a Cordin 550 high-speed digital camera was used in the experiments, synchronized with the loading stress pulses. For specimens with axes inclined to the SHPB loading axis, damage in the specimens will inevitably initiate from the stress-concentrated corners. The high-speed camera thus becomes a necessary tool in capturing the damage initiation point such that the local stresses can be calculated through numerical simulations using the corresponding stress level on the transmitted signals. Thirty two frames were captured for each experiment at a frame rate of 200,000 frames per second.

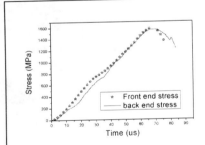

Figure 3. Stress history in the glass specimen with WC platens.

EXPERIMENTAL RESULTS

Effects of platen material: To study weather the effects of platen material on the crack initiation and propagation in the glass specimen, we used polished tool steel and WC platens. The experimental results show very similar stress-strain behavior in the specimens. The failure stress in the specimen is slightly higher when WC platens are used. Figure 3 shows the stress history of an experiment with a pair of WC platens. The stress history was measured both at the front and back end faces of the specimen, which shows a dynamically equilibrated stress state in the glass specimen. The specimen had a right prismatic geometry (no tilting angle). The stress histories shown in Fig. 3 show that the specimen was in near dynamic equilibrium under dynamic axial loading.

Figure 4 shows the dynamic damage and failure processes in the glass specimen that went through the stress history show in Fig. 3. In Fig. 4, the stress history is plotted again on the top of the figure. The numbers 1-4 marked on the stress-history curve corresponds to the instants when the high-speed images were taken and shown below in the figure.

Image #1 shows that cracks have been initiated from one end of the glass specimen. In image #2, as the earlier cracks propagate along the specimen axis, more cracks are initiated, both at the specimen/platen interface and inside the specimen. Both images # 1 and 2 were taken before the peak stress in the specimen is reached. Much more extensive damage is observed in images #3 and 4 as the stress in the specimen has gone beyond the compressive strength and the specimen starts to disintegrate. As will be discussed in more details later, the damage and failure processes in the glass

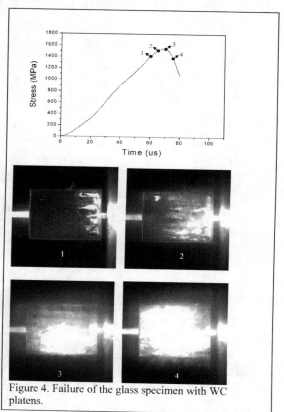

Figure 4. Failure of the glass specimen with WC platens.

specimen with steel platens showed very similar behavior, although the steel platens have a much higher Poisson's ratio. The reason is outlined in the following idealized analysis.

When an isotropic linear elastic material is under uniaxial loading, the lateral strain can be computed as

$$\varepsilon_{22} = \frac{1}{E}\left[\sigma_{22} - \nu(\sigma_{33} + \sigma_{11})\right] \tag{1}$$

where σ_{11} is the axial stress, σ_{22} and σ_{33} are lateral stress that are zero, E is the Young's modulus, and ν is the Poisson's ratio. Under uniaxial stress, the lateral strain is proportional to the axial stress σ_{11} through a constant $-\nu/E$.

The hardened and polished tool steel platens have a higher Poisson's ratio than that of the glass specimen (0.19). However, their Young's modulus is also higher than that of the glass (61 GPa). The WC platens have a comparable Poisson's ratio as the specimen but much higher Young's modulus. The constant $-\nu/E$ is 3.11×10^{-12} in the glass specimen, 1.45×10^{-12} in the steel platens, and 0.35×10^{-12} in the WC platens. Thus under axial loading, the lateral strains in both the steel and WC platens are less than that in the specimen. The WC platens constrain the specimen in the lateral direction more severe, resulting in slightly higher axial peak stress achieved in the specimen. However, the lateral strain in the steel platen is closer to that in the specimen, which may reveal more realistic mechanical behavior of the glass specimen under compression. We use the tool steel platens in the rest of the experiments reported in this paper to study the effects of shear stress.

Failure of a right prismatic specimen: Figure 5 shows a typical set of original oscilloscope recordings of the incident, reflected, and transmitted signals from a SHPB experiment on a glass specimen. The marked plateau on the reflected signal indicates that this specimen experienced a constant strain rate before failure occurred. Upon failure, the amplitude of the reflected signal increased drastically. This corresponds to the situation where the shattered specimen provides little resistance to the advance of the incident bar end. Figure 6 shows the dynamic stress histories on the incident and transmission bar ends, which indicates that the glass specimen was under dynamic equilibrium until the maximum strength was reached where the specimen became unstable. The experimental records shown in Figs. 5 and 6 suggest that, under dynamic equilibrium, the specimen in this experiment deformed at a nearly constant strain rate from the very early stage of deformation until very close to specimen final failure.

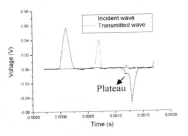

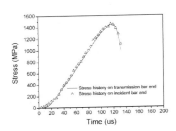

Figure 5. A set of SHPB experimental records.   Figure 6. Dynamic equilibrium in specimen.

The stress history in Fig. 6 shows that the ultimate compressive strength of the glass specimen reached 1.5 GPa, which is slightly lower than that reached in the experiments on the same materials with WC platens (Fig. 3).

Figure 7 shows the sequential high-speed images and the associated axial stress-time history obtained for a 0° specimen. The six numbered-points on the stress-time history curve correspond to the instants when the six images were taken. In the experiments, the flash lights were placed at the same side as the high-speed camera. With all the light traveling through the transparent glass specimen, the initial images appeared to be dark. When cracks are initiated, the cracked/damaged areas in a glass specimen can be easily seen in the optical images as bright regions since the crack surfaces are highly reflective to the incoming strobe light. A time interval of 5μs was set between any two consecutive images. Images prior to image # 1 did not show any feature that reflected light, revealing a uniformly deforming specimen. Images after image # 6 show extensively damaged/cracked stages of the specimen. The images in between, as shown in Fig. 7, reveal the time sequence of the damage initiation and propagation processes under the loading stress ramp. The circled area in frame # 2 contains a small crack very close to the end of the transmission bar, which indicates damage initiation in the specimen. In frame #3, nearly one third of the specimen is damaged by axial cracks emanating from the transmission bar-specimen end. However, another group of cracks can be seen to appear from the inside of the specimen (circled area). It is clear that, in addition to surface flaws, internal defected areas are also damage/crack

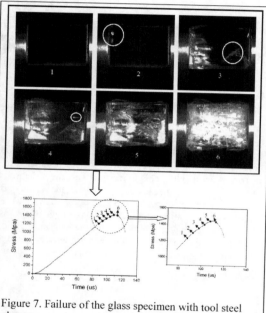

Figure 7. Failure of the glass specimen with tool steel platens.

initiators. This is particularly true for the 0° specimen since the geometry-induced stress concentration is kept minimum with this specimen configuration. In frame # 4 the cracks initiated from left side are propagating through the specimen in axial direction; and the internal cracks initiated in frame # 3 remain to be stationary. However, a new internal crack, as circled in frame # 4, appears just above the stationary group. Many more cracks are seen to have initiated and coalesced in frame # 5. Finally, the damage in the specimen is nearly saturated in frame # 6. During the damage propagation process, it is seen that the cracks, once initiated, propagate roughly along the specimen axis, which coincides with the SHPB loading axis for this specific case (0° specimen). Before any dominant crack can propagate through the entire specimen, other cracks initiate and propagate in the specimen. Some of the initiation sites are inside the specimen. Cracks initiated from the incident bar end (right side in the images in Fig. 7) finally coalesced with cracks initiated from the transmission bar end, leading to axial splitting and eventual crushing of the specimen. From the high-speed images shown in Fig. 7, we can estimate the axial cracks propagate at a speed of approximately 560 m/s.

Failure of a tilted prismatic specimen: To study the effects of introducing shear stress on the damage and failure behavior of the borosilicate glass, we conducted SHPB experiments with similar pulses described above on specimens with their axes inclined at 3°, 5° and 7° from the SHPB loading axis. These tilting angles introduce various degrees of shear stresses in the specimen. Results and related high-speed images from the 7° specimen are shown in Fig. 8 in the same format as in Fig. 7. As shown in the figure, damage/cracking in the titled specimens always initiate from the specimen corners with stress-concentration. The subsequent crack propagation into the specimen is roughly along the specimen axis, rather than along the SHPB compressive loading direction. The stress-time history curve of the angled specimen is also marked with the instants when the corresponding high-speed images were taken. A circled area indicates the initiation

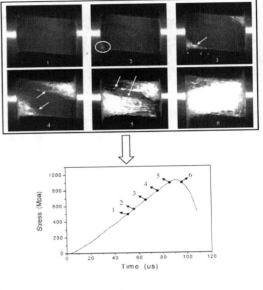

Figure 8. Failure of a glass specimen with a tilted specimen axis.

of damage at the corners. Time interval between points 1 and 2 in Fig. 8 is still 5 μs, whereas the interval thereafter is 10 μs. The reason for increasing the time interval is, as can be seen from the figures, the time duration from failure initiation to final specimen collapse increases with increasing specimen tilting angle. The damage initiation starts earlier with increasing tilting angle due to the stress concentration and the increased amount of shear stress at the corners. For an angled specimen, the stress distribution on the specimen/bar interfaces is no longer uniform. The stress in stress-time history curves shown in Fig. 8 is the average stress history captured by the strain gages mounted on the transmission bar. The local stress in the vicinity of the obtuse corner is considerably larger than the average stress in the specimen. Thus, damage always initiates from the obtuse angle area, which is verified in our observations as presented in Fig. 8. Furthermore, the observed average stress at failure decreases with increasing specimen angles (from 1302 MPa for 0° specimen to 560 MPa for 7° specimen). While the material fails in the stress concentration area, the other regions of the specimen are still in a relatively low stress state. Therefore, the damage initiates in the tilted specimens at lower average stress levels but the catastrophic failure of the entire specimen comes much later. It took only ~20 μs for a 0° specimen to experience the entire failure process. However, it took as long as 35 μs for a 7° specimen to go from damage initiation to catastrophic collapse. The measured maximum crack propagation speed is in the range of 600-700 m/s.

Failure initiation in a tilted prismatic specimen: The results shown in Fig. 8 illustrate that the onset of damage in the glass specimen can be accurately determined by the high-speed camera. This capability of damage initiation determination in angled specimens facilitates the determination of dynamic strength of the borosilicate glass under compression/shear loading. Stress concentration is inevitably introduced by the geometry of the tilted specimens. To identify the failure stress that corresponds to damage initiation in the specimen, local stress state at the corners, rather than average stress, needs to be used. However, the SHPB measurements provide only the averaged stress transmitted from the specimen. We used finite element analysis with a commercial code ABAQUS, together with the SHPB stress history data and the high-speed images to determine the stress state in the

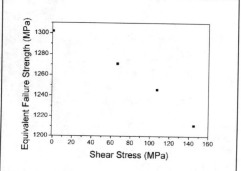

Figure 9. Effect of shear stress on the failure initiation in the glass specimen.

specimen when damage initiates. Figure 9 shows the trend that the failure strength, in terms of equivalent stress, of the borosilicate glass is sensitive to the imposed shear component. With increased shear, the equivalent stress at failure initiation decreases.

CONCLUSIONS

Borosilicate glass specimens with different tilting angles were dynamically loaded with a modified SHPB. The tilted specimen geometry introduces a shear component to axial compression, creating a principal stress state of axial compression and lateral tension. This facilitates the investigation of the effects of added shear on the dynamic failure process of the glass material at high rates of deformation. Linear ramp loading was generated through pulse shaping on the SHPB to load the linear material at a constant rate. A high-speed digital camera was used to record the damage initiation and failure process in the glass specimens. The images were synchronized with the loading history. Numerical stress analysis was used to determine the local stress state at the onset of damage at the obtuse corners in the specimens.

The local stress, in the form of both axial compressive component and the equivalent stress at the onset of damage was found to decrease with increasing shear stress. Cracks propagate in the specimen at a velocity in the range of 550-700 m/s. Cracks in specimens with or without tilting angles all propagate along the specimen axial direction, instead of the global SHPB loading axis. In the 0° specimens, cracks were found to initiate from both internal flaws and the bar/specimen interfaces.

REFERENCES

[1] N. S. Brar, S. J. Bless and Z. Rosenberg, "Impact-induced failure waves in glass bars and plates", Applied Physics Letters, 1991, Vol. 59, Issue 26, 3396-3398

[2] S. Sundaram and R. J. Clifton, "Flow Behavior of Soda-Lime glass at high pressures and high shear rates", Shock Compression of Condensed Matter, 1997, 517-520

[3] R. J. Clifton, M. Mello and N. S. Brar, "Effect of shear on failure waves in Soda Lime glass", Shock Compression of Condensed Matter, 1997, 521-524

[4] B. F. Raiser, J. L. Wise, R. J. Clifton, D. E. Grady and D. E. Cox, "Plate impact response of ceramics and glasses", Journal of Applied Physics, 75, 1994, 3862-3869

[5] N. Bourne, J. Millett, Z. Rosenberg and N. Murray, "On the shock induced failure of brittle solid", Journal of Mechanics and Physics of Solids, 46 (10), 1998, 1887-1908

[6] K. A. Dannemann, A. E. Nicholls, C. E. Anderson Jr., I. S. Chocron and J. D. Walker, "Compression Testing and Response of Borosilicate Glass: Intact and Damaged", Proceedings of Advanced Ceramics and Composites Conference, January 23-27, 2006 Cocoa Beach, FL

[7] G. T. Gray, "Classic Split Hopkinson Pressure Bar Technique,"ASM Handbook, Vol. 8, Mechanical Testing and Evaluation, ASM International, Materials Park, OH (2000)

[8] H. Kolsky, "An investigation of the mechanical properties of materials at very high rates of loading", Proceedings of the Royal Society of London, 1949, B62, 676-700

BALLISTIC PERFORMANCE OF COMMERCIALLY AVAILABLE SAINT-GOBAIN SAPPHIRE TRANSPARENT ARMOR COMPOSITES

Christopher D. Jones, Jeffrey B. Rioux, and John W. Locher
Saint-Gobain Crystals
33 Powers Street
Milford, NH, 03055

Vincent Pluen, and Mattias Mandelartz
Saint-Gobain Sully
16 route d'Isdes
Sully sur Loire, France

ABSTRACT

Saint-Gobain has developed a ceramic transparent armor composite using our Edge-defined Film-fed Growth (EFG) CLASS™ Sapphire crystals to defeat various ballistic projectiles. The high strength, high hardness, and good transparency in the visible and infrared spectra make sapphire an ideal choice for transparent armor applications. The composites are commercially available in large sizes (greater than 300 mm wide and 775 mm long) with substantial weight and thickness improvements over classical laminated glass solutions. This paper will discuss the properties and benefits of using Saint-Gobain sapphire transparent armor composites, along with the ballistic results for certain thicknesses and areal densities for defeating both single shot and multi-hit ballistic results of armor piercing rounds.

INTRODUCTION

It has been one year since we first introduced our sapphire transparent armor system to the public, and during this year we have emerged as the sole, fully integrated commercial supplier of ceramic transparent armor systems.[1,2] The combination of expertise in Saint-Gobain Crystals (ceramic materials) and Saint-Gobain Sully (lamination and ballistics) has resulted in products that lead the field with respect to ballistic performance and weight savings. Indeed, Saint-Gobain Crystals has been growing large sapphire sheets in production since 2003, growing thousands of crystals proving the commercial viability of large transparent ceramic materials.[3-9] Today, Saint-Gobain Crystals offers their CLASS[225™] and CLASS[300™] sapphire in production quantities, see Figure 1. The largest application of these windows, in addition to transparent armor, has been in the aerospace industry for VIS-MWIR optical windows, where the intrinsic strength, hardness (resistance to erosion), chemical resistance, and high optical transmission make sapphire a highly desirable material. Of course, these same properties make sapphire a highly desirable material for transparent armor.

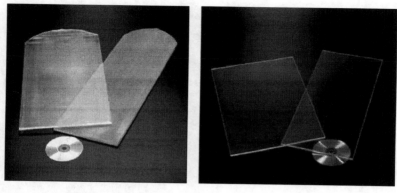

Figure 1. CLASS$^{300^{TM}}$ and CLASS$^{225^{TM}}$ sheets as grown (left) and after polishing (right).

Traditional transparent armor systems use laminated glass with a polymer backing to stop ballistic projectiles. These systems can be quite thick and heavy due to the amount of glass needed to stop high powered projectiles. As threats become tougher to defeat and specific threats become harder to stop with traditional laminated glass, ceramic composites are the leading solution to reduce thickness and weight, while improving protection and transparency. The four leading strike-face materials used in armor are glass, sapphire (Al_2O_3), aluminum oxynitride [$AlN_x(Al_2O_3)_{1-x}$], and spinel ($MgAl_2O_4$). Figure 2 contains pertinent material properties for each of these materials.[10-12] Sapphire has the best fracture toughness, flexure strength, thermal conductivity, hardness and modulus of elasticity (Young's modulus) of all the strike face materials. The density of sapphire is almost double glass, but only slightly higher than the other ceramic-based strike-face materials. Based on the mechanical properties, sapphire is expected to have the best ballistic performance of all the strike-face materials for transparent armor.

BALLISTIC TESTING

The sapphire transparent armor composites are composed of a CLASS™ sapphire sheet on the strike face, see Figure 3, and the sapphire is bonded to glass using conventional interlayers, such as polyvinyl butyral. While the exact interlayer and composition of each window is proprietary, all samples have the CLASS™ sapphire strike face, interlayers of glass, and a final polymer layer on the backside of the window.

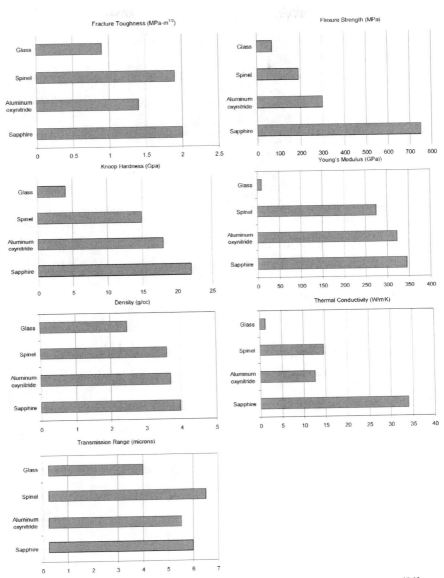

Figure 2. Selected material properties of glass, spinel, aluminum oxynitride and sapphire[10-12]

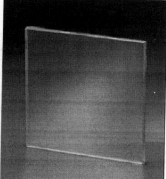

Figure 3: 305 x 305 mm sapphire transparent armor system. The system is composed on CLASS™ sapphire bonded to glass and polymers.

Each sapphire transparent armor composite was tested by either a single shot in the middle of a 150 x 150 mm sample ("single shot test") or by multiple hits in a triangle (~120 mm spacing) of a 305 x 305 mm sample ("multi-hit test"), see Figure 4. Any deviation to this testing procedure is noted in the text. Due to the existence of multiple test standards, we will not report on meeting specific standards (although, Saint-Gobain can provide this for commercial orders), but rather the projectiles and velocities within each test. These results are summarized in Table I.

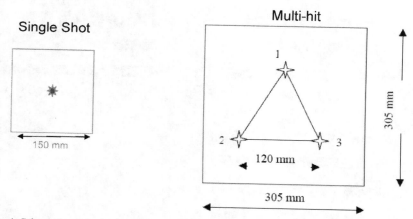

Figure 4: Schematic of single shot and multi-hit shot test samples.

Table I: Summary of ballistic tests using Sapphire Transparent Armor. ID# 1-8 are single shot with ID 9-11 being multi-hit.

ID	Sample Size (mm)	Thickness (mm)	Areal Density kg/m2	Projectile	Projectile Velocity m/s	Penetration
1	150	21.1	52.12	7.62x51 M-80 Ball	835	Partial
2	150	21.1	52.12	7.62x39 API-BZ	776	Partial
3	150	21.1	52.12	7.62x51 AP(M61)	768	Partial
4	150	21.1	52.12	7.62x51 AP(M61)	846	Complete
5	150	29.4	72.78	7.62x54R B32	857	Partial
6	150	24.87	67.46	7.62x54R B32	864	Partial
7	150	43.78	159.21	7.62x51 AP-WC	917	Partial
8	150	46.3	169.11	7.62x51 AP-WC	921	Partial
9	305	41.85	107.97	7.62 x 63mm APM2	Shot 1 Shot 2 Shot 3	Partial Partial Partial
10	305	20.78	55.6	7.62x39 API-BZ	Shot 1:715.9 Shot 2: 712.7 Shot 3: 718.8	Partial Partial Partial
11	305	24.78	65.6	7.62 x 51 AP(M61)	Shot 1:842.6 Shot 2: 837.8 Shot 3: 837.2	Partial Partial Complete

7.62x51mm M-80 Ball (lead ball round).

It was previously reported that the 7.62x51mm M-80 ball was defeated using a sapphire armor system that was 21.1 mm thick and had an areal density of 52.12 kg/m^2, see Figure 5.[1,2] The projectile velocity was 835 m/s for the single shot test and there was no perforation of the polymer backside.

Figure 5: Sapphire transparent armor system (150x150 mm) defeating single shot 7.62x51mm M-80 Ball.

7.62x39mm API-BZ (armor piercing with hardened steel core).

It was previously reported that it was possible to defeat the single shot 7.62x39mm API-BZ at velocities up to 776 m/s with a sapphire transparent armor system at 21.1 mm thick and 52.12 kg/m², see Figure 6.[1,2] We have improved on this make-up and now report that it is possible to defeat the 7.62x39mm API-BZ multi-hit with a thinner make-up 20.78 mm. The sapphire transparent armor system after the multi-hit test is shown in Figure 7.

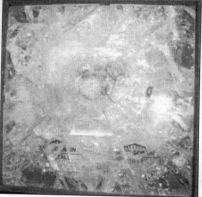 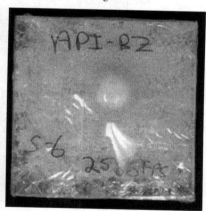

Figure 6: Sapphire transparent armor system (150x150 mm) defeating single shot 7.62x39mm API-BZ

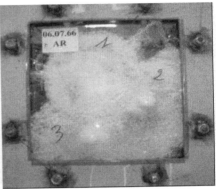

Figure 7: Sapphire transparent armor system (305x305 mm) defeating multi-hit 7.62x39mm API-BZ

7.62x54R B32 (armor piercing with hardened steel core).

It was previously reported that it was possible to stop a single shot 7.62x54R B32 at a velocity of 857 m/s with a sapphire transparent armor system at 29.4 mm thick and a areal density of 72.78 kg/m². [1,2] We have improved on the previous composition to show that it is now possible to stop the 7.62x54R B32 with a sapphire transparent armor system at 24.87 mm thick with an areal density of 67.46 kg/m², see Figure 8. The armor piercing core of this ballistic projectile is shown in Figure 9 before and after testing, showing that the sapphire strike face destroys the projectile, and thus reducing the armor piercing capabilities.

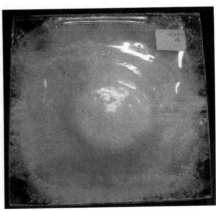

Figure 8: Sapphire transparent armor system (150x150 mm) defeating single shot 7.62x54R B32

Figure 9: 7.62x54R B32 API hard projectile core before (left) and after impact (right) on sapphire strike face. Note the total destruction of the hard projectile core after contact with the sapphire transparent armor system.

7.62x51 AP-WC (AP8) (armor piercing with tungsten carbide hard core).

Two compositions were shot with this high-powered ballistic round. The 7.62x51 AP-WC was defeated at 917 m/s with a sapphire transparent armor system 43.78mm thick with an areal density of 159.21 kg/m^2. The second composition also defeated this round at 921 m/s with a thickness of 46.3 mm and an areal density of 169.11 kg/m^2, see Figure 10.

Figure 10: Sapphire transparent armor system (150x150 mm) defeating single shot 7.62x51 AP-WC AP8

7.62 x 63 APM2 (armor piercing with hardened steel core).

It is reported that the it is possible to defeat the 7.62x63 APM2 multi-hit test at 41.85 mm thick with an areal density of 107.97 kg/m². The sapphire transparent armor system after testing is shown in Figure 11. It should be noted that there was no perforation of the final polymer layer.

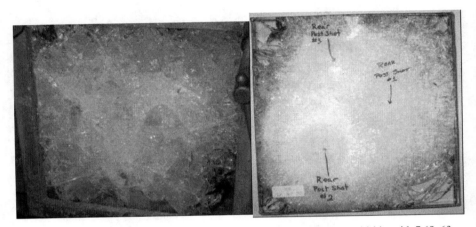

Figure 11: Sapphire transparent armor system (305 x 305 mm) defeating multi-hits with 7.62x63 APM2

7.62x51 AP (M61) (armor piercing with hardened steel core).

It was previously reported that it was possible to defeat the 7.62x51 AP(M61) at a velocity of 776 m/s with a sapphire transparent armor system 21.1 mm thick with an areal density of 52.12 kg/m².[1,2] This velocity was slightly slower than normal testing velocities (830 m/s), which showed complete penetration at this thickness and areal density. A new sapphire transparent armor composition was tested at 24.78 mm with an areal density of 65.6 kg/m² which was capable of defeating 2/3 multi-hit shots (Figure 12). The first two ballistic projectiles were stopped, however the last shot penetrated the final polymer layer.

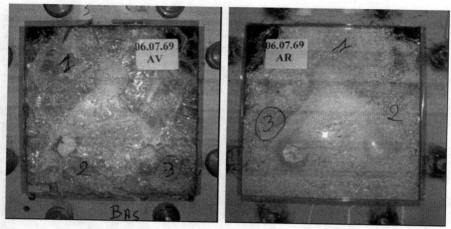

Figure 12: Sapphire transparent armor system (305 x 305 mm) defeating 2 of 3 shots in multi-hits with 7.62x51 AP

Optical Tests

A significant advantage of using sapphire as a strike-face is the high optical transmission in the visible and infrared spectrum. A sapphire transparent armor system at 29.4 mm thickness that can defeat the 7.62x54R B32 round has a transmission greater than 85%, with haze levels around 1%, see Table II. Measurements were performed according to ASTM method D1003. For comparison, a typical glass window that could defeat the 7.62x54R B32 round in glass would be 55 mm thick with a luminous transmission of 73% and haze around 0.6%, showing the vast improvement in transmission by using a sapphire transparent armor system. Increasing the thickness of the system by adding additional glass layers to 41.1mm thickness only slightly reduces the transmission to around 84%, with no change in haze.

Table II: Optical Measurements on CLASS™ sapphire transparent armor system

Thickness (mm)	Areal Density (kg/m^2)	Luminous Transmission (%)	Haze (%)
29.4	72.78	85.9	0.99
29.4	72.78	85.3	1.25
41.1	101.11	84.4	1.00
41.1	101.11	84.3	1.12

Comparison of Sapphire Transparent Armor to Traditional Glass Armor

As we have demonstrated above, sapphire transparent armor can defeat a variety of threats, including multi-hit rounds. The advantage of using sapphire as a strike face is that it has high flexure strength, high hardness, and good toughness (see Figure 2). These properties contribute to the destruction of the hard core in armor piercing projectiles, thus greatly reducing the armor piercing capabilities of the threat (see Figure 9). Table III shows the comparison between sapphire transparent armor and traditional glass armor to defeat two different armor piercing rounds. To defeat a single shot 7.62x54R B32 API, it is possible to have a thickness savings of 55% and a weight savings of 41% by using a sapphire transparent armor system. To defeat a multi-hit 7.62x39 API BZ, it is possible to realize thickness savings of 64% and weight savings of 58%.

Table III: Comparison of Sapphire Transparent armor to Glass armor

Threat	# shots	Glass Thickness (mm)	Glass Areal Density (kg/m2)	Sapphire Armor Thickness (mm)	Sapphire Armor Areal Density (kg/m2)	Thickness Savings with Sapphire Armor	Weight savings with sapphire armor
7.62x54R B32 API	1	55	115	24.8	67.5	55%	41%
7.62x39 API BZ	3	58	133	20.78	55.6	64%	58%

Commercialization

The commercialization of ceramic armor is dependent on a variety of factors, including availability, quality, cost, and type of the ceramic strike face, the glass type, interlayers, construction, and the final polymer layer. Saint-Gobain has the unique capabilities to address this market and continue to be the leading supplier of ceramic transparent armor. The commercial availability of good quality CLASS™ sapphire strike faces up to 305 mm wide and 775 mm long at reasonable cost has enabled Saint-Gobain to gain market-share where other ceramic materials are neither available in this size in commercial quantities nor good quality. The expertise within Saint-Gobain in understanding the construction, lamination, and ballistic testing has also proved to be critical assets in helping customers define a sapphire transparent armor that can meet their needs.

CONCLUSION

Saint-Gobain has developed a commercial ceramic transparent armor system based on CLASS™ sapphire. The sapphire transparent armor system can improve the thickness and weight, by up to 64% and 58%, respectively, over traditional glass transparent armor. Additionally, improvements in window transmission can be realized by using a sapphire transparent armor system over a traditional glass system. This is due in part to the excellent mechanical, physical, and optical properties of sapphire as a strike-face material. The thickness and areal density of systems to defeat single shot 7.62x51 M80 Ball, 7.62x39 API-BZ, 7.62x51 AP(M61), 7.62x54R B32, 7.62x51 AP WC (AP8) and multi-hit 7.62xAPI-BZ, 7.62x51 AP(M61), and 7.62x63 APM2 are presented.

REFERENCES

1. C. D. Jones, J. B. Rioux, J. W. Locher, H. E. Bates, S. Zanella, V. Pluen, M. Mandelartz, Large-Area Sapphire for Transparent Armor, Proceedings of the 30th International Conference on Advanced Ceramics and Composites, The American Ceramic Society, January 2006.

2. C. D. Jones, J. B. Rioux, J. W. Locher, H. E. Bates, S. Zanella, V. Pluen, M. Mandelartz, Large-Area Sapphire for Transparent Armor, Amer. Cer. Soc. Bull., Vol. 85, No. 3 March 2006.

3. H. E. LaBelle, *EFG the invention and application to sapphire growth*, Journal of Crystal Growth, Vol. 50, 8 (1980).

4. J.W. Locher, H.E. Bates, S.A. Zanella, E.C. Lundstedt, and C.T. Warner, *The production of 225 x 325 mm sapphire windows for IR (1 to5 μm) application*, Proc. SPIE Vol. 5078, Window and Dome Technologies VIII. Randal W. Tustison, Editor, 2003, pp.40-46

5. J. W. Locher, J. B. Rioux, and H. E. Bates, *Refractive Index Homogeneity of EFG A-plane Sapphire for Aerospace Windows Applications*, 10th DoD Electromagnetic Windows Symposium, Norfolk VA, 18-20 May 2004.

6. H. E. Bates, C. D. Jones, and J.W. Locher, *Optical and crystalline characteristics of large EFG Sapphire Sheet*, Proc. SPIE Vol. 5786, Window and Dome Technologies and Materials IX, Randal W. Tustison, Editor, May 2005, pp. 165-174

7. J. W. Locher, H.E. Bates, C. D. Jones, and S. A. Zanella, *Producing large EFG sapphire sheet for VIS-IR (500-5000 nm) window applications*, Proc. SPIE Vol. 5786, Window and Dome Technologies and Materials IX, Randal W. Tustison, Editor, May 2005, pp. 147-153

8. C. D. Jones, J. W. Locher, H.E. Bates, and S. A. Zanella, *Producing large EFG sapphire sheet for VIS-IR (500-5000 nm) window applications*, Proc. SPIE Volume 5990, 599007 (2005).

9. V. A. Tatartchenko, *Sapphire Crystal Growth and Applications*, in Bulk Crystal Growth of Electronic, Optical, and Optoelectronic Materials, edited by P. Crapper, pp 299-338, John Wiley & Sons (2005)

10. D.C. Harris, *Materials for infrared windows and domes, properties and performance*, SPIE Optical Engineering Press, Bellingham WA, 36 (1999).

11. Surmet Corporation Company Literature #M301030; 2004

12. Saint-Gobain Crystals Product Literature; 2004

Opaque Ceramics

SYNTHESIS OF CERAMIC EUTECTICS USING MICROWAVE PROCESSING

Anton V. Polotai, Jiping Cheng, Dinesh K. Agrawal, Elizabeth C. Dickey*
and Sheldon Cytron**

Materials Research Institute, the Pennsylvania State University, University Park, PA, 16802
* Department of Materials Science and Engineering, the Pennsylvania State University,
University Park, PA, 16802
** U.S. Army TACOM-ARDEC, Picatinny, NJ 07806

ABSTRACT

This communication presents preliminary results of using microwave energy in order to melt and re-solidify oxide and non-oxide refractory ceramic eutectic compositions in an appropriate crucible. The $Al_2O_3 - Y_3Al_5O_{12}$ (YAG) ($T_m = 1827°C$) and B_4C-TiB_2 ($T_m = 2310°C$) eutectic compositions were melted in a multimode 2.45GHz microwave furnace using a specially designed insulating package based on boron nitride. Our experiments demonstrate the stability of eutectic melt temperature (the absence of temperature runaway) and the uniformity of subsequent eutectic microstructure across the sample diameter during the re-solidification of big ~7cm^3 samples. The ability to reach ultra-high temperatures (T ~ 2400°C) and prevent the formation of significant thermal gradient across the sample confirms the potential of using microwave energy for processing of directionally solidified refractory ceramic eutectic compositions.

INTRODUCTION

Directionally solidified eutectics (DSEs) of oxide and non-oxide ceramic compositions are attractive composite materials due to their unique thermodynamic, mechanical and electrical properties[1]. In general these materials have excellent thermal stability, high-temperature strength and fracture toughness, which make them attractive candidates for ultra-high temperature structural materials[2]. In addition to their outstanding mechanical properties, some of the DSE compositions of rare-earth, alkali-earth and d-transition metal borides possess other exceptional properties such as high-electron emission, high neutron absorption ability, and specific magnetic and electrical characteristics[3].

The low thermal gradient across a sample and good temperature control during the re-solidification are critical factors for processing the high-quality eutectic rods with a diameter higher than 1 centimeter. Currently, there are several methods for fabricating DSEs, which can be divided into two main classes based on their process heat propagation scheme. The first class incorporates methods in which the heat is generated by an external source and propagates from the surface to the center of the ceramic rod. This class includes heating in a conventional resistive furnace[4], inductive heating with an external susceptor[5], infrared heating by halogen or xenon lamps[6], laser beam heating[7], and heating by electric arc or by electron beam bombardment[8]. In most instances, the ceramic rod is a sintered rod of ceramic powders of the desired eutectic composition. The main drawback of all of these methods is the presence of a thermal gradient within the rod, which may lead to an inhomogeneity of microstructure and restrict the sample diameter. The other class of melting techniques incorporates methods in which the heat is generated directly inside the sample, minimizing the formation of thermal gradients and potentially allowing for scaling to larger sample rods. One of the representatives of this class is inductive heating of conductive materials without an external susceptor[9]. In spite of the internal heating advantage, this method can not be utilized for fabricating broad classes of non-conducting ceramic eutectics because it is suitable only for conductors in which eddy currents can be generated.

Another source of heat generation in this class is microwave heating capable to work with non-conductive dielectric materials, while maintaining the same advantages as inductive heating[10]. The microwave heating of ceramics is based on creating internal friction of the atoms/molecules or defects in the microstructure of the materials located in a microwave field[10]. The penetration or skin depth of microwave in dielectric materials is given by

$$D = \frac{3.22 * 10^7}{f\sqrt{\varepsilon}\tan\delta} \tag{1}$$

in which f is microwave frequency, ε is dielectric constant and $\tan\delta$ is called the loss tangent[10]. For example, the dielectric loss and the half power depth of alumina, quartz and mullite are 0.009, 0.0004. 0.00054 and 4500 mm, 67800mm, 69450 mm at 2.45 GHz, respectively. So, the microwave energy penetration depth depends on the dielectric loss of the material and generally is much longer than the diameter of processed samples ensuring the uniform heating of the sample. Very little heating occurs when microwaves are reflected by the material (such as bulk metals), or fully penetrate through the material (such as perfect insulators).

Microwave heating has found broad applicability in different branches of industry replacing some of the more conventional heating technologies[10]. However, its use for the fabrication of oxide and non-oxide ceramic eutectics has not been explored. One of the primary issues in microwave processing this class of materials is identifying effective thermal insulating materials capable of working at ultra-high temperatures without coupling with the microwave energy. Here, we report the first results on achieving ultra-high temperatures as high as 2400°C and uniform heating using microwave energy to melt oxide and non-oxide eutectic compositions in an appropriate crucible assembly.

EXPERIMENTAL PROCEDURE

For the melting experiments the eutectic compositions with 81.7 mol% of Al_2O_3 – 18.3 mol% of Y_2O_3 (T_{eut} = 1827°C)[12] and 88 mol% of B_4C – 12 mol% of TiB_2 (T_{eut} = 2310°C)[13] have been selected. High purity commercial powders: α-Al_2O_3 (99.99%, ~1μm. Alfa Aesar, Ward Hill, MA), Y_2O_3 (99.99%, <10μm, Alfa Aesar, Ward Hill, MA). B_4C (99.4%, 1-7μm, Alfa Aesar, Ward Hill, MA) and TiB_2 (99.5%, ~40μm, Alfa Aesar, Ward Hill, MA) were used in this study. The Al_2O_3-Y_2O_3 and B_4C-TiB_2 powders were ball milled in a polyethylene jar with yttria-stabilized zirconia milling media in ethanol for 20 hours to obtain homogeneously mixed powder compositions. The obtained slurries were dried in a rotary evaporator to remove the ethanol. Green pellets with dimensions of 20 mm in diameter and 25 mm in height were pressed in a steel die with 77 MPa of uniaxial pressure followed by cold isostatic pressing at 200 MPa. The sample pellets were subsequently pre-sintered in a conventional furnace at 1600°C for 2 hours in air for the Al_2O_3-YAG composition and in pure Ar atmosphere for the B_4C-TiB_2 composition.

The microwave melting experiments were conducted by using a laboratory made 6 kW, 2.45 GHz multimode microwave cavity. As most refractory materials change their dielectric properties at high and ultra-high temperatures and start to couple microwave energy, special attention was paid to the design of proper sample insulation packages. A schematic drawing of such a package is presented in Fig. 1.

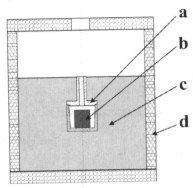

Figure 1. The schematic of configuration of the insulation package for microwave heating of the eutectics: (a) Boron nitride crucible, (b) sample, (c) boron nitride powder, (d) Fiberfrax[tm] thermal insulator.

Boron nitride was chosen as a main insulation material for its inertness to most metallic and salt melts, superior high-temperature stability and transparency to MW energy up to ultra-high temperatures higher than 2200°C[14]. The insulation package consisted of the following parts: BN crucible with a lid and cone-shaped internal space, which was machined from a hot-pressed HBT-type BN rod (GE Advanced Materials, Strongsville, OH), BN powder (99.5%, ~40μm, Alfa Aesar, Ward Hill, MA), which provided at least 30 mm thick thermal insulation shield between BN crucible and the thin external thermal insulation package (Fiberfrax[tm], Unifrax, Niagara Falls, NY). The atmosphere inside the microwave chamber was ultra-high purity argon under ambient pressure. To prevent plasma formation inside the crucible, the sample was embedded in powder of the same composition. Since the Al_2O_3-Y_2O_3 composition is very good insulator material with low dielectric losses, it is transparent to microwaves at room temperature. A thin layer of carbon powder was painted on the external surface of the BN crucible as the susceptor to initiate heating. The temperature was monitored using infrared pyrometers (Raytek MA2SC Raytek, Santa Cruz, CA) (working temperature range 350-2000°C) and (Mirage Series OR15-35C, IRCON Infrared Thermometers, Niles, IL) (working temperature range 1500-3500°C) and recorded in situ by a computer along with forwarded and reflected power data from the microwave generator. After the melt appeared, 10 minutes of isothermal holding was maintained in both cases. Cooled samples were cut in two by a high-speed diamond saw and were examined by scanning electron microscopy (SEM, Hitachi SEM-S-3500N, Instruments Group, Hitachi Ltd., Tokyo, Japan).

RESULTS AND DISCUSSION
The general temperature-power-time profiles of the melting experiments of Al_2O_3-YAG and B_4C-TiB_2 composites are presented in Fig. 2. At the beginning, the temperature of the Al_2O_3-YAG sample climbed rapidly with a heating rate up to 120°C/min. Subsequent heating up to 1500°C resulted in a gradual decrease of the heating rate. When the temperature had reached ~1500°C the heating rate rapidly increased up to 140°C/min until a melt appeared. The increase of heating rate at temperatures higher than 1500°C is associated with the initiation of significant

coupling of microwave energy by alumina[11]. The appearance of a liquid phase was accompanied by a noticeable decrease in reflected power (marked with a circle and an ellipse on the graphs), and heating rate. The temperature was found to be mostly constant during 10 minutes after the melt appeared. This result is expected because the energy is going toward the enthalpy of melting rather than raising temperature. In the case of the B_4C-TiB_2 composition, the heating rate was more or less steady and much faster (~260°C/min) over the whole temperature range up to the melting point compared to the Al_2O_3-YAG. Similar to the oxide composition, after the melt appeared the temperature was stable and almost independent of the forwarded microwave power. The decrease of reflected power during powder melting and constant melt temperature are important for the material processing because, firstly, it allows an alternative way to monitor the melt appearance and, secondly, it allows the constant melt temperature, which is a necessary condition for the successful directional solidification.

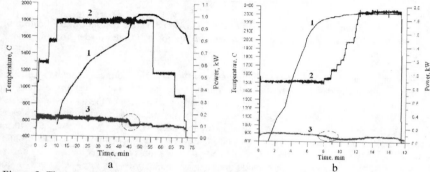

a   b

Figure 2. The temperature-power-time profiles of the experiments during microwave melting of Al_2O_3-YAG (a) and B_4C-TiB_2 (b) eutectics. Lines 1, 2 and 3 correspond to the temperature, forwarded power and reflective power, respectively.

The cross-sectional view of processed samples and general microstructures of both oxide and non-oxide eutectics are presented in Figures 3 and 4. The Al_2O_3-YAG composition shows a typical cellular eutectic microstructure[2,12] (fig. 3), in which the dark phase is Al_2O_3 and the white phase is YAG. In spite of the significant sample volume ~7cm³, the microstructure of Al_2O_3-YAG eutectic was almost the same at outer edges and in the center of sample suggesting the existence of uniform composition and temperature gradient during re-solidification process. The B_4C-TiB_2 sample shows some deviation from correct eutectic composition toward to the B_4C compound (fig. 4). The eutectic grains consist of the dark B_4C matrix phase with white elongated TiB_2 lamellae. In spite of the B_4C phase excess, the B_4C-TiB_2 sample also shows the uniformity of eutectic grain microstructure along the sample diameter. Since the solidification process was not directional in our experiment, the B_4C-TiB_2 eutectic grains were aligned randomly. The stable melt temperature and the formation of uniform eutectic microstructure inside the big refractory samples confirms that the microwave heating could be suitable for the directional solidification of big diameter oxide and non-oxide high-melting point eutectic rods. In all of the experiments the boron nitride crucible survived at ultra-high temperature under microwave irradiation and showed no visible reaction with the melted eutectics. The melted samples did not adhere to the crucible walls and were easily removed after cooling.

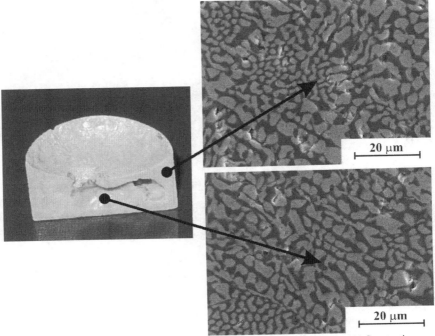

Figure 3. The cross section view and corresponding microstructure of Al_2O_3-YAG eutectic composition.

All melted samples exhibited the formation of some large bubbles and cavities inside the sample body. The spherical cavities suggest that they were formed by extended vapor pressure due to the existence of some dissolved gases in the melt or decomposition of the chemical compounds. In eutectic processing within a crucible, such gaseous pores may be eliminated by allowing extra time and higher melt temperature to stabilize the eutectic or by lowering the atmospheric pressure to allow gas diffusion out of the sample. However, this may work only with low vapor pressure compounds such as oxides. In non-crucible directional solidification processes, such microstructural inhomogeneity is generally eliminated by mixing the liquid pool through rotation of feed and seed rods in opposite directions.

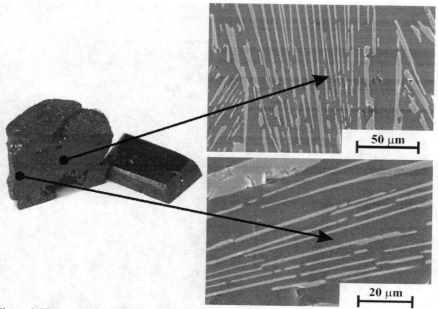

Figure 4. The cross section view and corresponding microstructure of TiB_2-B_4C eutectic composition.

CONCLUSIONS AND FUTURE WORK

The Al_2O_3-YAG and B_4C-TiB_2 eutectic compositions were melted in a 2.45 GHz multimode microwave furnace using a specially designed thermal insulation package. Boron nitride was used successfully as the crucible and primary thermal isolation material. The eutectics were heated up to 2350°C and melted in an inert atmosphere. It was found that the sample temperature was very stable when the melt appeared, and the subsequent eutectic microstructure was uniform across sample. These studies indicate the potential use of microwave energy for directional solidification of high-melting point eutectic compositions. The microwave energy eliminates significant thermal gradient across the processed rod, allowing one to process much bigger eutectic rods. Our ongoing work in this area is aimed at designing a unique microwave furnace for directional solidification studies of refractory eutectics.

ACKNOWLEDGMENTS

Supported by the U.S. Department of the Army, under grant No. W15QKN-04-D-1012.

REFERENCES

[1]R. L. Ashbrook, "Directionally Solidified Ceramic Eutectics," J. Am. Ceram. Soc., 60 (9-10) 428-435 (1977).

[2]See Papers of the Directionally Solidified Eutectic Meeting, Paris, France, 03-05 May 2003 in J. Europ. Ceram. Soc., 25(8) 1191-1462 (2005). Edited by A. Sayir, M.-H. Berger, J. Fuller and Y. Waku.

[3]Yu. B. Paderno, V. N. Paderno, V. B. Filippov, Yu. V. Mil'man and A.N. Martynenko, "Structure Features of the Eutectic Alloys of Borides with the d- and f- Transition Metals," Soviet Powder Metallurgy and Metal Ceramics, 31 8(356) 700-706 (1993).

[4]D. Viechnicki and F. Schmid, "Growth of Large Monocrystals of Al_2O_3 by a Gradient Furnace Technique," J. Crystal Growth, 11 [3] 345-347 (1971).

[5]Y. Waku, N. Nakagawa, T. Wakamoto, H. Ohtsubo, K. Shimizu and Y. Kohtoku, "Excellent High-Temperature Properties of YAG Matrix Composites Reinforced with Sapphire Phases," Processing and Fabrication of Advanced Materials IV, 323-339 (1996).

[6]I. Gunjishima, T. Akashi and T. Goto, "Characterization of Directionally Solidified B_4C-TiB_2 Composites Prepared by Floating Zone Method," Mater. Trans., 43 [4] 712-720 (2002).

[7]I. de Francisco, R. I. Merino, V. M. Orera, A. Larrea and J. I. Pena, "Growth of Al_2O_3/ZrO_2 (Y_2O_3) Eutectic Rods by the Laser Floating Zone Technique: Effect of the Rotation," J. Europ. Ceram. Soc., 25 1341-1350 (2005).

[8]C. M. Chen, L. T. Zhang and W. C. Zhou, "Characterization of LaB_6-ZrB_2 Eutectic Composite Grown by the Floating Zone Method," J. Crystal Growth, 191 873-878 (1998).

[9]C. C. Sorrel, "The Directional Solidification and Properties of the ZrC-ZrB_2 and ZrC-TiB_2 Eutectics," MS thesis, the Pennsylvania State University (1980).

[10]E. Kubel, "Special Focus: Advancements in Microwave Heating Technology," Industrial Heating, 1 43-53 (2005).

[11]M. A. Janney, H. D. Kimrey, "Microwave Sintering of Alumina at 28 GHz," Ceramic Powder Science II. 1 919-924 (1998).

[12]D. Viechnicki and F. Schmid, "Eutectic Solidification in the System Al_2O_3/$Y_3Al_5O_{12}$," J. Mater. Sci., 4 84-88 (1969).

[13]S. S. Ordan'yan, E. K. Stepanenko, A. I. Dmitriev and M. V. Shchemeleva, "Interaction in the B_4C-TiB_2 System," Soviet Journal of Superhard Materials, (English Translation of Sverkhtverdye Materialy), 8 (5) 34-37 (1986).

[14]C. E. Holcombe, N. L. Dykes, "Importance of "Casketing" for Microwave sintering of materials," J. Mater. Sci. Letters, 9 425-428 (1990).

HOT PRESSING OF BORON CARBIDE USING METALLIC CARBIDES AS ADDITIVES

Pedro Augusto de S.L. Cosentino, D.Sc, Major Military Engineer
Materials Laboratory – Brazilian Army Technological Center
28705 Américas Av
Rio de Janeiro/RJ – BRAZIL

Célio A. Costa, Ph.D
Federal University of Rio de Janeiro - Technological Center – Metall and Mat Engng Dept
Universitary City
Rio de Janeiro/RJ - BRAZIL

José Brant de Campos, D.Sc.
National Institute of Technology
82 Venezuela Av
Rio de Janeiro/RJ – BRAZIL

Roberto Ribeiro de Avillez, Ph.D
Catholic University of Rio de Janeiro - Metall and Mat Sci Dept
225 Marquês de São Vicente St
Gávea – Rio de Janeiro/RJ – BRAZIL

ABSTRACT

The effect of metallic carbides (chromium and vanadium carbide) as additives for boron carbide sintering and the particle size distribution were analyzed in this study. Carbon was also used as standard additive. Particle size was changed by high energy milling, using planetary mill for two and four hours. The as received and the two hours milled were hot pressed at 1800°C, pressure of 20 MPa and argon atmosphere, with additives. The densities of sintered samples varied form 80% to 99% to as received and two hours milled powder, respectively. The analysis showed the presence of new phases at the grain boundary for the two hours milled powder, which kept the grain size small during the sintering process.

INTRODUCTION

The advanced ceramic field is constantly searching for new processing technologies and chemical compositions, which can lead to improvements in physical and mechanical properties. Processing itself is a very critical step, since it deals from powder purity and particle size distribution to sintering, phases present and grain size [1, 2]. Furthermore, most of the processing conditions are done far from the equilibrium [3].

Boron carbide (B_4C) is a structural ceramic material, with a relative small world-wide production [4]. The most used process to manufacture solid pieces of B_4C is by solid state sintering [5]. It is worth mention that sintering techniques are effective in producing high density structural ceramics, that is, near 100% of the theoretical density of the material.

However, boron carbide sintering is too difficult, requiring specials furnaces which use temperatures over 2000° C, controlled atmosphere and pressure. The origin of these difficulties of

B_4C sintering arises from the covalent bonds, which allow the mass transport mechanisms (volume and grain boundary diffusion) and eliminate pores, to operate only near the melt temperature [4]. To reduce sintering temperature, the use of pressure, additives and reductions of particle size have been applied [4]. The use of small amount of carbon, which is the standard additive, is necessary for sintering [4].

Such conditions make the material processing expensive and, consequently, limit its applications. However, B_4C has some strategic applications such as light armor, control bars for nuclear reactors and components subject to high abrasion [5,6,7]. For instance, in armor applications, where over 95% of theoretical density is required, sintering by hot pressing (HP) or hot isostatic pressing (HIP) in temperatures above 2000°C is quite common [4].

To further reduce sintering temperature or to eliminate pressure, research has focused on new sintering additives such as metals, oxides and carbides [8-13] and reduction of the particle size [14]. Attention has to be given to new phases formed, since they may be harder or softer than B_4C, and this can harm the densification process and/or hardness of the material [15].

In this study the effect of particle size reduction and the effect of sintering additives (VC and Cr_3C_2) on B_4C densification will be demonstrated – samples with carbon were also sintered. Basically, the mechanochemical activation by the high energy milling process of particle size reduction allied to the metallic carbides as additives showed a synergic effect which favored densification.

MATERIALS AND METHODS

The as received materials used in this work are listed in Table 1.

Table 1 – The as received materials and their origin.

Material	Formula	Manufacturer	Use
Boron carbide	B_4C	ESK	Matrix
Vanadium carbide	VC	HCStarck	Addittive
Chromium carbide	Cr_3C_2	HCStarck	Addittive
Carbon	C	CTEx	Addittive

To reduce the average particle size of the as received B_4C a high energy milling process was carried out using a planetary mill (RETSCH PM 4) for two and four hours. The steel vessel was coated with WC and the comminution process was carried out with zirconia spheres (0.4 to 0.7 mm) in isopropyl alcohol (IPA). Due to the extreme hardness and abrasiveness of the boron carbide, zirconia micron particles and other elements of the sintering vessels and coating were incorporated into the mixture in small amounts and were considered as additives too. After milling, zirconia spheres were separated form the pulp, which was then dried in a oven.

A mixture of the milling powder and the additives C (B_4C+C), VC (B_4C+VC) and Cr_3C_2 ($B_4C+Cr_3C_2$) were homogenized with IPA and alumina spheres for 12 hours. The amounts of each additive were 2 and 4 % in weight. After drying, the samples were hot pressed under 20 MPa and sintered at 1800°C for 2 hours, under argon atmosphere.

After HP, the pellets were ground and polished with diamond, and analyzed by the following techniques: Archimedes's method, Helium picnometry, X-ray diffraction. The microstructural analysis were done with Scanning Electron Microscope (SEM - JEOL JSM-6460 LV) allied with the Energy Dispersive Spectroscopy (EDS).

The combination of SEM and the X-ray diffraction (Rietveld method) were used to individually identify the phases formed during sintering. For thew XRD was used the X`Pert Pan-Analítical PRO, with CuK$_\alpha$ radiation, scanning step of 0.05° with a collection time of 5 seconds.

The Rietveld method parameters, suggested by Cheary and Coelho [16], to identify and quantify precisely the phases formed, were used here and the data were analyzed using the software Bruker-AXS TOPAS 2.1.

RESULTS AND DISCUSSION

The particle size distribution indices of the as received and the milled powder are shown in Table 2, where it can be observed that two and four hour milling times had the same effect in particle size distribution. Such behavior was corroborated by BET analysis, which showed the same surface area for milling B$_4$C for two and four hours. The milling process used here has been reported to result in nano particles of zirconia (from the milling spheres) incorporated to the milled mass [14, 17, 18].

Table 2 – The particle size distribution indices

Milling time	d$_{10}$	d$_{50}$	d$_{90}$	d$_{100}$
As received	0.20	0.42	2.89	12.21
2h	0.20	0.36	1.71	3.60
4h	0.20	0.37	1.89	3.60

Since the comminution process resulted pretty much in the same powder, samples were hot pressed with the as-received and 2h milled powder the mixture. The sintered samples had the density measured by Archimedes and inferred by X-ray diffraction, and the phases present were analyzed by X-ray diffraction combined with the Rietveld method [14].

The density measurements are shown in Figure 1, where it can be observed that the sintered samples which used the as received powder densified to about 84% of the theoretical density measured by the Rietveld method; on the other hand, the sintered samples which used the milled powder densified to almost 99% of the theoretical density measured by the same Rietveld method. Furthermore, the crystallographic density was increased to about 2.8 g/cm^3, whilst means a severe contaminated by the milling process and, also, through sintering heavy phases were generated.

To achieve density at near theoretical is extremely difficult in B$_4$C. Usually, high density requires small grains that keep themselves small during the sintering process, since they favor the mass transport in lower temperatures [19]. An effective way to control grain growth was the observed precipitation at the grain boundary as shown in Figure 2 and by the phases identified in X-ray diffraction (Figure 3).

However, the effectiveness of keeping the grain small depends on the removal of B$_2$O$_3$ from the surface of the B$_4$C next to 1800°C; otherwise, high density will not be achieved [19, 20, 21].

The presence of oxides on the surface of the as received and milled powder were not quantified, but the high energy milling combined with the metallic carbide additives likely favored by the elimination of the residual B_2O_3 from the surface of the powder. It also limited grain growth by precipitation as reported by other authors [8, 10, 12].

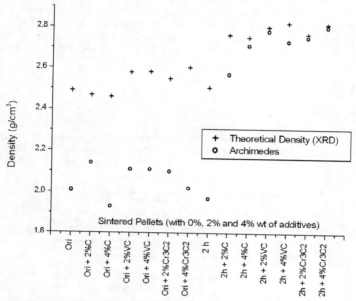

Figure 1 – Theoretical density (obtained from XRD data) and Archimedes density (g/cm³) of the sintered pellets. "Ori" means the pellets were made with original powder without milling. "2h" mean the pellets were made with 2 hours milled powder.

The hot pressed samples of the 2h milled powder showed higher density than the HP sample of the as received powder. The Rietveld analysis was used to explain these differences. The mixture with metallic carbides and the hot press used here promoted the reaction of finely dispersed particles, forming new phases in the matrix located at the grain boundaries, as shown in Figure 2.

When the X-ray diffraction of the hot pressed sample with 4% Cr_3C_2 is analyzed (Figure 3), it is observed a variety of new phases formed and the amount of $B_{13}C_2$ was reduced to 84.4%, a significant decrease in the main phase when compared to Figure 4, which is a typical X-ray diffraction pattern of B_4C hot pressed with the original powder, where the only phases present are $B_{13}C_2$ in amount of 98.1% and graphite.

The origin and the variation, in terms of %$B_{13}C_2$ and %C, was effect of additive use, milling process and the singular crystalline structure of the $B_{13}C_2$, which is an arrangement of icosahedrons of $B_{11}C$ distorted located in the vertices of a rombohedral Bravais cell (space group

R3m) [22]. This combination was able to change the chemical equilibrium between boron and carbon in the crystalline structure.

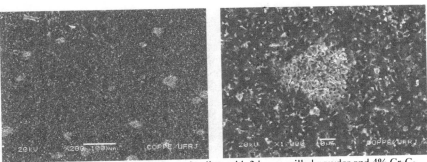

Figure 2 – Microstructure of the sintered pellets with 2 hours milled powder and 4% Cr_3C_2 additives (200x e 1k x).

The X-ray diffraction identified the main phase as $B_{13}C_2$, which corresponds to icosahedron of $B_{11}C$ linked with C-B-B (carbon-boron-boron) chains. Once the chemical equilibrium was altered, boron or carbon can move out form C-B-B chain and form a new phase outside of the icosahedron. The free carbon/boron could react with both the metallic carbide additives and the contaminants from milling process and form the new phases observed in Figure 3. Such phases are heavier than the original $B_{13}C_2$, which explain the higher crystallographic density measured.

The mixture with metallic carbides and the hot press used here promoted the reaction of finely dispersed particles, forming new phases in the matrix located at the grain boundaries, as shown in Figure 2.

The formation of these new phases were responsible for the increase in the measured density observed and impair the sintering process. The Rietveld method combined with MEV/EDS identified and quantified the constituent phases of the sintered pellets.

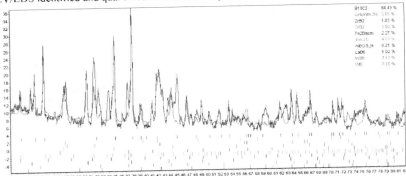

Figure 3 – Difratogram (blue) and Rietveld refinement (red) of the B_4C pellet with 2 hours milled powder and 4% Cr_3C_2 additive.

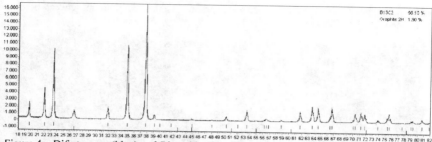

Figure 4 – Difratogram (blue) and Rietveld refinement (red) of the sample hot pressed with the as received powder.

CONCLUSIONS

From the analyzed data the main conclusions are the following:

- The use of a planetary mill skewed the particle size distribution, but contamination was introduced in the system.
- The hot press temperature used (1800°C) was lower than the usual temperature of boron carbide sintering, which is of the order of 2100°C;
- Densities of about 99% were obtained with the 2 h milled powder;
- When milled powder has to be used, the determination of the real density is very important, and the Rietveld method was shown to be satisfactory;
- The samples hot pressed with the as received powder reached a relative density of 80% at most, which showed that the particle size distribution was more relevant than the average particle size or the additives used.

BIBLIOGRAPHY

1. Kingery, W.D., Introduction of Ceramics, John Wiley and Sons, Inc., p. 16, 1976.
2. Barsoum, M., Fundamentals of Ceramics, McGraw Hill, 1997, p. 10.
3. Suryanarayana, C., "Mechanical Alloying and Milling", Progress in Materials Science 46, pp 1 – 184, 2001.
4. Thévenot, F., "Boron Carbide – A Comprehensive Review", J of the European Ceramic Society, vol. 6, pp.205-225, 1990.
5. Lipp, A. "Boron Carbide – Production, Properties, Application", Technische Rundschau, Numbers 14, 28, 33 (1965) e 7 (1966), Elektroschmetzwerk Kempten Gmbh, München.
6. Matchen, B. "Applications of Ceramic in Armor Products", Key Engng Mat, vols. 122-124, pp 333-342, 1996.
7. Ceradyne Advanced Technical Products, "Complete Protection Solutions for 21st Century Warfare", 2004.
8. Sigl, L.S., "Processing and Mechanical Properties of Boron Carbide Sintered with TiC", J of the European Ceramic Society, vol. 18, pp.1521-1529, 1998.

9.   Rogl, P. e Bittermann, H., "Ternary Metal Boron Carbides", Int J of Refractory & Hard Materials, vol. 17, pp 27-32, 1999.

10.   Radev, D.D. e Zakhariev, Z., "Structural and Mechanical Properties of Activated Sintered Boron Carbide basedf Materials", J of Solid State Chemistry, vol. 137, pp 1-5, 1998.

11.   Melo, F.C., "Adding Effects on Boron Carbide Sintering", Ph.D. Thesis, IPEN/USP, Brazil, 1994.

12.   Deng, J. et al., "Microstructure and Mechanical Properties of Hot Pressed $B_4C/(W,Ti)C$ Ceramic Composites", Ceramics International, vol. 28, pp 425-430, 2000.

13.   Levin, L. e Frage, N., "The Effect of Ti and TiO_2 Additions on the Pressureless Sintering of B_4C", Metallurgical and Mat Trans A, vol 30A, p 3201, Dez 1999.

14.   Cosentino, P.A.S.L., "Metallic Carbides Additives for Boron Carbide Sintering Using Hot Pressing", Ph.D. Thesis, COPEE/UFRJ, Brazil, 2006.

15.   McCauley, J. et al. "Microstructural Characterization of Commercial Hot-Pressed Boron Carbide Ceramics", J. Am. Ceram. Soc., 88 [7] 1935–1942, 2005.

16.   Cheary, R.W. e Coelho, A., "A fundamental parameters approach to X-Ray line -profile fitting", J. Appl. Cryst., 25, pp. 109-121, 1992.

17.   Santos, M.A.P., "Processing and sintering of national silicon carbide", Ph.D. Thesis, COPPE/UFRJ, Brazil, 2003.

18.   Rawle, A., "Basic Principles of Particle Size Analysis", Technical Paper, Malvern Instruments Limited.

19.   Prochazka, S. e Dole, S.L., "Microstructural coarsening during sintering of boron carbide", J of the American Ceramic Society, 72, 6, pp 958 - 966, 1989.

20.   Lee, H. e Speyer, R.F., "Advances in presureless densification of boron carbide", J of Materials Science, 39, pp 6017-6021, 2004.

21.   Schwetz, K.A. e Grellner, W., "The influence of carbon on the microstructure and mechanical properties of sintered boron carbide", J of Less Common Metals, 82, pp 37-47, 1982.

22.   Lazzari R. et al, "Atomic structure and vibrational properties of icosaedral B_4C boron carbide", Physical Review Letters, 83, 16, pp 3230 - 3233, 1999.

ACKNOWLEDGEMENTS

The authors would like to acknowledge the Materials Division of the Aerospace Technology General Command of the Brazilian Air Force (AMR/CTA/FAB) and specially the contribution of Dr Francisco "Pires" Cristovão Lourenço de Melo and his help with the hot press tests.

SPATIAL DISTRIBUTION OF DEFECTS IN SILICON CARBIDE AND ITS CORRELATION WITH LOCALIZED PROPERTY MEASUREMENTS

Memduh V. Demirbas, Richard A. Haber, Raymond E. Brennan
Rutgers University
Ceramic and Materials Engineering
607 Taylor Road
Piscataway, NJ 08854

ABSTRACT

Spatial distributions of defects, mainly pores and inclusions, could be a critical component in armor performance, although, this has been not addressed sufficiently. Characterization of defects was performed on three hot-pressed silicon carbide plates by scanning electron microscopy, nondestructive evaluation by ultrasound and image analysis. Nearest neighbor distance distributions were used to perform spatial data analysis, quantify microstructures and help to determine how pores were spatially distributed on several polished sections of armor grade silicon carbide. Localized property measurements were implemented using Knoop indentation as a tool to correlate between microstructural data and quasi-static properties. Different loads were used in order to change the effective area of an indent and Weibull modulus values were obtained for each load. As a continuation to microstructural analysis part, spatial data analysis was performed on ultrasound images and a correlation between microstructural data and ultrasound data was obtained.

INTRODUCTION

Ceramics must be free of any unwanted inclusions and possess high theoretical density with small amounts of porosity in order to live up to their potential superior properties. Carbonaceous defects are commonly present in ceramic armor samples [1] and porosity in small percentages is also unavoidable, therefore, more thorough characterization on these two types of flaws is required.

The spatial arrangement of pores or second phases is a microstructural characteristic expected to affect the ballistic performance of ceramic armor materials. Therefore, it is of interest to estimate the statistical descriptors of the spatial arrangement of porosity and inclusions. Nearest neighbor distance distributions will serve as an important tool to examine spatial distributions and the corresponding mean nearest neighbor distances of defects [2].

The results obtained from microstructural analyses should be validated using quasi-static or dynamic tests. Hardness tests were chosen for this purpose. Viechnicki *et al.* claimed that it is the only static mechanical property that alone helps predict ballistic performance [3]. Although contradictory studies were done [4], at least there is an agreement in armor community that hardness is a key parameter in terms of ballistic performance. [3, 5, 6]

Nondestructive evaluation of materials has been an important tool for testing materials since no damage is made during examination and it is also a method frequently used for assessing armor materials [7]. Critical defects that are not easily detected within the bulk could be identified using nondestructive evaluation (NDE) by ultrasound, which will be employed in this study to correlate with microstructural assessment. The data from C-scan imaging will be used for performing spatial data analysis.

EXPERIMENTAL PROCEDURE

Three commercial hot-pressed SiC samples from a single production lot were used throughout this study. Two of the samples were under a minimum density and considered "rejects" and one tile passed the inspection. Rejected tile 1 had a low density region and rejected tile 2 had a white spot on the surface. The sample with the low density region will be referred to as "Low density"; the one with the white spot will be referred to as "Defect" and the accepted tile will be referred to as "Armor grade".

Indentations were performed on randomly selected regions of the samples. The distance between each indent was chosen as 0.5 mm. The indents were spaced far apart so that they did not interfere with each other. Testing started at 2 Kg and went down using lower loads at 1 Kg, 0.5 Kg, 0.3 Kg and 0.1 Kg. A range of indentation loads were used in order to examine the effect of local variations of defects on hardness results with the change in indent size. Weibull plots were used to examine the spread in the data and to test reliability in terms of hardness. This analysis was also applied in order to get a representative number, Weibull modulus, out of the mechanical tests so that a correlation between microstructural assessments and property measurements could be obtained.

Nearest neighbor distance distributions were employed for the microstructural assessment part. By using this method, the distance between a pore and its closest neighbor was calculated. This was performed for all pores or other features in the field of view. A distribution of nearest neighbor distances was obtained at the end [8,9]. The image analysis procedure and more background information on nearest neighbor distance distributions were explained thoroughly in previously published papers by Demirbas and Haber [10,11].

Ideally, a rapid nondestructive means of assessing microstructures is needed. Therefore, an important component will be nondestructive evaluation by ultrasound. Destructive testing is not always indicative of the performance of the remaining untested plates from a production lot. Anomalous or critical defects that are not easily detected within the bulk and could prove to be detrimental to the performance of the armor plates could be present. Ultrasound and microscopy will be coupled in examining those defects within the bulk in this paper.

For the ultrasound study, tiles were scanned using a 125 MHz transducer to provide a high degree of detail and resolution for detection of defects. Higher frequency transducers produce ultrasound waves with shorter wavelengths that enable the detection of smaller features in the test sample. This increases the probability that a small defect will be detected. However, one drawback is that the degree of attenuation, or loss of ultrasound energy, increases at higher frequencies. Since attenuation is also a function of the thickness and homogeneity of the sample, the optimum transducer frequency must be chosen based on the desired level of detection and the integrity of the sample. Consequently, as long as the reflected ultrasound signals from the material can be resolved, higher frequencies are preferred [12].

The amplitude range was approximately between 0 and 45 mV for the C-scan images and the scans were separated into nine different ranges before a "threshold" filter was applied. Threshold filter assigns 1 to features and 0 to background so the image becomes binary. Some data points would have been included in the background after "threshold" filter was applied if the separation had not taken place and information would have been lost. The overall amplitude range was broken down into several ranges in order to avoid loss of information.

An image analysis procedure was developed for evaluation of C-scan images. First, a "threshold" filter was applied on the images. Then, they were inverted to reveal the low density regions, which were shown with white color before inversion. A "dilate" filter is used to reveal the features and a "watershed" filter helped separate the connecting ones. A "fill holes" filter was applied to get rid of the open regions inside features. Finally, a "cut off" filter was applied to remove noise in the images.

RESULTS

Microstructural Assessment

The micrographs from the three samples can be seen in Figure 1. On the appearance, they have similar average pore size and spatial distribution, however the level of clustering of porosity cannot be ascertained. It must be pointed out that some of these defects accepted as pores are actually carbon pullouts so a spatial characterization of carbon inclusions is also inherent in this study.

Nearest neighbor distance distributions were performed on these samples using image analysis and the results can be seen in Figure 2. The 'More' column in the plot refers to values larger than 15 μm for the nearest neighbor distance distributions. "Armor grade" and "Defect" show relatively narrow nearest neighbor distance distributions with the variance values of 3.47 and 2.94 μm^2, respectively. This narrow distribution of nearest neighbor distances (nnd) can be observed qualitatively on the right side of the graph, as "Low density" has higher peaks above 10 μm. "Low density" shows a higher variance value with 5.51 μm^2 and broader distribution compared to the other two samples. This is an indication of inhomogeneous distribution of porosity that "Low density" possesses.

Use of two parameters, Q and V, was suggested by Anson and Gruzlezski [8] to be more quantitative in terms of spatial distributions, where;

$$Q = \frac{\mu_o}{\mu_e} = \frac{\text{observed mean nearest - neighbor distances}}{\text{expected mean nearest - neighbor distances}}$$

$$V = \frac{\text{var}_o}{\text{var}_e} = \frac{\text{observed variance of nearest - neighbor distances}}{\text{expected variance of nearest - neighbor distances}}$$

the expected mean and variance are the expected mean and variance for a random distribution. Q-V plots for these three samples are given in Figure 3. Different types of spatial distributions can be categorized by the following guidelines: (a) random distributions, Q ≈ 1 and V ≈ 1; (b) regular distributions, Q > 1 and V < 1; (c) clustered distribution, Q < 1 and V < 1; and (d) random distribution with clusters, Q < 1 and V > 1 [13].

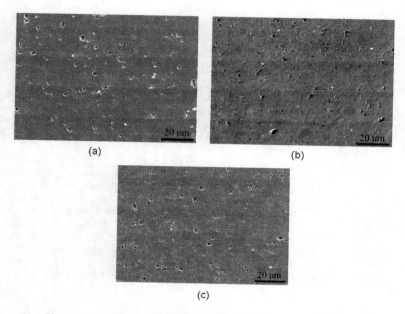

Figure 1. SEM micrographs of (a) Low density (b) Defect (c) Armor grade

The region in between two lines by the y-axis also represents "random" distribution. All three samples fall in between the two lines, which indicate random distribution for all of them. However, "Low density" is in the region of random distribution with clusters showing a subtle difference from "Armor grade" and "Defect".

Property Measurements

The later part of this study was to determine whether local property variations correlated with the microstructural assessment. As mentioned previously, hardness was chosen for that purpose. Hardness data was obtained by using Knoop indenter and five different loads. being 2 Kg. 1 Kg. 0.5 Kg. 0.3 Kg and 0.1 Kg. to see the change in the effective area of indents.

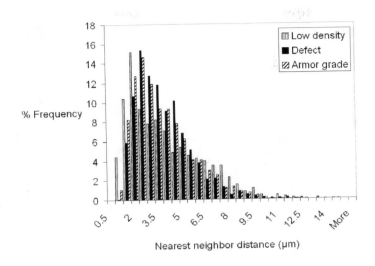

Figure 2. Nearest neighbor distance distributions of hot-pressed SiC samples

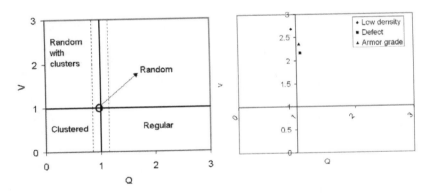

Figure 3. (a) Q-V plot showing each spatial distribution condition (b) Q-V plot for the hot-pressed SiC samples

One hundred indents were placed on each sample in order to have statistically meaningful data for further evaluation.

Weibull analysis was performed on the hardness data and the spread in the Weibull plots was examined [14.15].

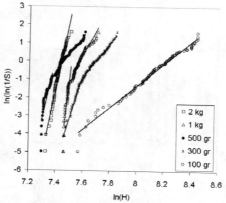

Figure 4. Weibull plots for "Low density" at five different loads

The Weibull plots for each sample is given in Figures 4, 5 and 6 and each graph provides five plots from five different loads.

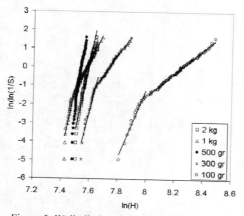

Figure 5. Weibull plots for "Defect" at five different loads

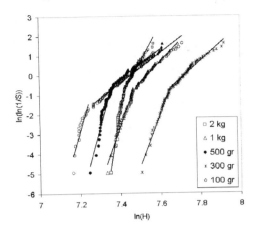

Figure 6. Weibull plots for "Armor grade" at five different loads

Multimodal Weibull distributions can be observed in the plots for all three samples and for most of the indentation loads except for 2 Kg and 0.1 Kg for "Low density" and 2 Kg and 0.5 Kg for "Defect". All five curves for "Armor grade" show bimodal distribution. Weibull moduli values, m_1 and m_2, are obtained from each plot and they are tabulated in Table 1. Different m values were given for different parts of the plots due to the significant change in the slopes and this change was determined by finding the point at which R^2 values started to decrease.

None of the samples stands out as the one with highest modulus values for all loads. "Armor grade" has the highest Weibull modulus with 59.5 for 2 Kg load. The trend is completely reversed for 1 Kg, with "Low density" having 67.9. "Defect" has the highest value for 0.5 Kg with 57.3 while "Low density" and "Armor grade" have very close modulus values with each, at 34.3 and 35.7, respectively. For 0.3 Kg, "Defect" possesses the highest value again with 32.5 while "Armor grade" has the largest Weibull modulus with 33.4 for 0.1 Kg. According to these results, there is no correlation between hardness values for different loads.

One important observation from these plots is that as the indentation loads decreases, the slopes of the curves tend to decrease accordingly. This observation is quantified where Weibull modulus values are plotted against the indentation load in Figure 6. Curve fits of m_2 values for all samples are high with R^2 values of 0.97, 0.89 and 0.90, respectively. The fits of m_1 values are not that good especially for "Defect" and "Low density" due to one of the modulus values for each one that does not fit the pattern.

The relationship between load and Weibull modulus values is an issue that should be considered. It can be speculated that as the indent size decreases, the hardness value depends very much on the number of pores under the indent since a few pores that interact with the indent might affect the results considerably. However, with higher indentation loads, the effective area under the indent is much larger; therefore, a small change in the number of defects that are in the vicinity of the indent does not affect the results significantly. Higher modulus values at higher loads could be explained using this thought process.

Grain size could be another issue that might affect the hardness results, but no significant difference of hardness results were observed among the three samples. The fact that the average grain size for all three samples is roughly 2 μm, cancels the possible effect of grain size on hardness for these samples.

Table I. Weibull modulus values for all samples at five different loads

		2 kg	1 kg	0.5 kg	0.3 kg	0.1 kg
"Low Density"	m_1	37.8	67.9	34.3	25.1	6.1
	m_2	-	13.5	9.7	9.2	-
"Defect"	m_1	38.9	36.7	57.3	32.5	18.7
	m_2	-	15.1	-	10.9	6.2
"Armor grade"	m_1	59.5	27.7	35.7	26.4	33.4
	m_2	18.5	10.4	7.4	9.2	7.6

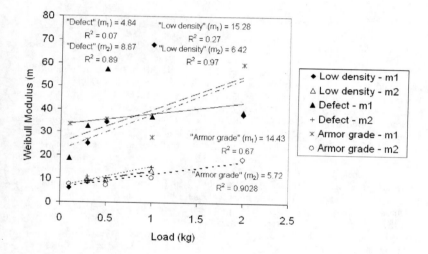

Figure 6. Weibull modulus values vs. indentation load to show the decrease in modulus values with decrease in load

Comparison between Microstructure and Property Results

It was shown in the microstructural analysis that "Armor grade" and "Defect" show more of random distribution as opposed to the clusters of pores superimposed on a random background that LD possesses. This was quantified by obtaining Q-V values and examining the nearest neighbor distance distributions.

Hardness results show mixed results with each indentation load. The trend from the highest hardness to the lowest changes for each load, however, the convention for looking at SiC armor plates using Knoop indenter is to use 2 kg load. Therefore, it is of value to compare results from 2 kg indentation load to the microstructural results. AG has two modes for 2 kg load, where m_1 is 59.5, much higher than those for "Low density" and "Defect", showing 37.8 and 38.9, respectively. The results from the microstructural parameters, Q and V, are shown in Figure 3 and as it can be observed, "Low density" point falls into the region of clusters superimposed on a random background. The other two samples do not show any sign of clustering. The distance of each point to the random point, (1, 1), can be calculated by using the following equation, $((Q-1)^2+(V-1)^2)^{0.5}$, and the overall ranking among the samples can be obtained. The calculated distances are 1.69, 1.17, and 1.36 for "Low density", "Defect", and "Armor grade", respectively. The smaller number represents a microstructure closer to the random distribution. "Low density" has the largest distance and the smallest Weibull modulus so microstructure results are in line for "Low density". However, "Defect" and "Armor grade" results from these analyses do not match up in terms of ranking among the samples. There is a significant variation in hardness results, which might contribute to this difference between the two types of analyses.

Spatial Analysis of Ultrasound C-Scan images

As mentioned previously, these three hot-pressed samples were tested nondestructively using ultrasound. Complete ultrasound analyses results are being compiled and will be published elsewhere. In this paper, the spatial analysis on C-scan images will be emphasized. The C-scans obtained at 125 MHz are given in Figure 7.

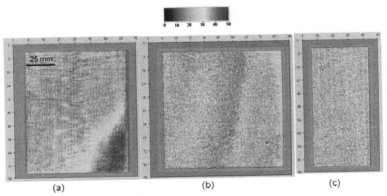

(a) (b) (c)

Figure 7. Ultrasound C-scans of, (a) "Low density" (b) "Defect" (c) "Armor grade", at 125 MHz

Although the images appear to be two dimensional, C-scan images are obtained from the signals that pass through the samples. Therefore, the results are collective of what is in the material throughout its thickness.

"Low density" name for that sample comes from the region of low density on the bottom. "Defect" has a single defect in the right corner and it has some striations in the middle parts of the scan while "Armor grade" has a uniform C-scan.

C-scan images are then processed by image analysis as explained previously. The images from which nearest neighbor distances are calculated is given in Figure 8.

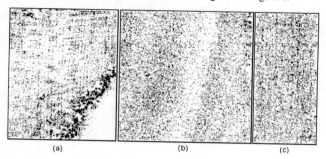

| (a) | (b) | (c) |

Figure 8. An example of image processed C-scan maps of (a) "Low density"
(b) "Defect" (c) "Armor grade"

The scans were separated in to nine different ranges and image analysis was performed on each one. The reason for this separation was not to lose any information on after a "threshold" filter and do this process in small ranges to include any related information.

Nearest neighbor distance distributions are calculated from the images at nine amplitude ranges. The results from 23-24 mV are given in Figure 9. The amount of points that fall into this range is 22.0% for "Low density", 21.0% for "Defect", and 20.8% for "Armor grade" of all amplitude values. The reason that this range is being reported here is that it covers the most of the ultrasound amplitude range among the separated ranges.

A large spread is observable for "Low density" where a significant number of data points extend over the given range. "Defect" shows a tail in its distribution graph but the tail is not as pronounced as it is for "Low density". "Armor grade" demonstrates a narrower distribution compared to the other two where largest nearest neighbor distance does not go beyond 0.09 inches. Three other curves that belong to the different data ranges show similar trend. Out of all the other eight amplitude ranges, only one shows "Defect" with the broadest nearest neighbor distance distribution curve. The results from the rest of the data ranges show similar distribution curves among each other.

Microscopy provides information from a single section inside of the material, while ultrasound looks at bulk properties. Therefore, there is a significant scale difference between the two techniques. This should be pointed out for correct interpretation of the results. Still, it is useful to obtain some degree of correlation between two methods for evaluation of SiC armor.

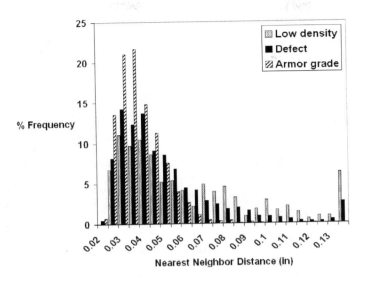

Figure 9. Nearest neighbor distance distributions at 23-24 mV

SUMMARY

Three commercial hot-pressed SiC samples from a single production lot were examined by spatial data analysis. Nearest neighbor distance distributions were employed for microstructural assessment part from the pictures obtained by using SEM. In this method, the distance between a pore and its closest neighbor is calculated and this is performed for all pores or other features in the field of view. The differences in the microstructures of samples were revealed as "Armor grade" and "Defect" showed microstructures closer to random distribution than "Low density".

Hardness tests were conducted at five different indentation loads and Weibull analysis was performed using the large number of indentation data. Signs of correlation between modulus values and the indentation load were observed, where Weibull moduli tend to decrease as the indentation load decreases.

Amplitude C-scan images by ultrasound were used for spatial assessment of defects. Ultrasound data was examined from the standpoint of "location". The nearest neighbor distance distributions from micrographs and from ultrasound results showed similar trends, and this could be an indication that both destructive and nondestructive testing point to similar tendencies.

ACKNOWLEDGEMENTS
The authors would like to thank NSF/IUCRC Program, Ceramic and Composite Materials Center (CCMC) and U.S. Army Research Laboratory Materials Center of Excellence (MCOE) for funding this research.

REFERENCES
[1] M.P. Bakas, V. A. Greenhut, D.E. Niesz, G. D. Quinn, J. W. McCauley, A. A. Wereszczak and J. J. Swab, "Anomalous defects and dynamic failure of armor ceramics", International Journal of Applied Ceramic Technology 1 (3), 211-218 (2004)
[2] A.Tewari and A. M. Gokhale, "Nearest neighbor distances in uniaxial fiber composites", Computational Materials Science, 31 (2004) 13-23
[3] D. Viechnicki, W. Blumenthal, M. Slavin, C. Tracy and H. Skeele, "Armor Ceramics – 1987", The Third Tacom Armor Coordinating Conference Proceedings
[4] P. Lundberg and B. Lundberg, "Transition between interface defeat and penetration for tungsten projectiles and four silicon carbide materials", International Journal of Impact Engineering 31 (2005) 781–792
[5] J. Sternberg, "Material Properties Determining the Resistance of Ceramics to High Velocity Penetration", J. Appl. Phys., Vol 65(9), 1 May 1989
[6] E. Medvedovski, "Alumina ceramics for ballistic protection – Part 2", American Ceramic Society Bulletin, Vol. 81, No. 4 (2002) pp. 45-50
[7] R. Brennan, R. Haber, D. Niesz and J. McCauley, "Defect Engineering of Samples for Non-Destructive Evaluation (NDE) Ultrasound Testing", Ceramic Armor and Armor Systems II: Proceedings of the 107th Annual Meeting of The American Ceramic Society, Baltimore, Maryland, USA 2005, Ceramic Transactions, Volume 178
[8] J. P. Anson and J. E. Gruzlezski, "The Quantitative Discrimination between Shrinkage and Gas Microporosity in Cast Aluminum Alloys Using Spatial Data Analysis" Materials Characterization 43, 319-335 (1999)
[9] J. Ohser and F. Mucklich, "Statistical Analysis of Microstructures in Materials Science", John Wiley & Sons, Ltd (2000)
[10] M.V. Demirbas and R. A. Haber, "Relationship of Microstructure and Hardness for Al_2O_3 Armor Material", Advances in Ceramic Armor II, CESP, p. 167-179, 2006
[11] M. V. Demirbas and R. A. Haber, "Defining Microstructural Tolerance Limits of Defects for SiC Armor", Ceramic Armor and Armor Systems II, Ceramic Transactions, 178, p. 109-122, 2005
[12] R.E. Brennan, R. Haber, D. Niesz, J. McCauley and M. Bhardwaj, "Non-Destructive Evaluation (NDE) of Ceramic Armor: Fundamentals", Advances in Ceramic Armor, 26 (7) pp.223-230 (2005)
[13] P. A. Karnezis, G. Durrant and B. Cantor, "Characterization of Reinforcement Distribution in Cast Al-Alloy/SiC_p Composites" Materials Characterization 40, 97-109 (1998)
[14] C. K. Lin and C. C. Berndt, "Statictical analysis of microhardness variations in thermal spray coatings", Journal of Materials Science 30 (1995) 111-117
[15] Jianfeng Li and Chuanxian Ding, "Determining microhardness and elastic modulus of plasma-sprayed Cr_3C_2-NiCr coatings using Knoop indentation testing", Surface and Coatings Technology 135 (2001) 229-237

SILICON CARBIDE MICROSTRUCTURE IMPROVEMENTS FOR ARMOR APPLICATIONS

Steven Mercurio, Richard A. Haber
Rutgers University
607 Taylor Rd.
Piscataway, NJ 08854

ABSTRACT

Modified processing methods and additives allow for the development of SiC with reduced grain size microstructures and improved mechanical properties. A surfactant based additive system for carbon and boron was coupled with particulate aluminum nitride additions to modify grain boundary behavior and grain growth. Near theoretical densities were attained at lower hot pressing temperatures ranging between 2000°C and 2200°C. Microscopic evaluation indicates a decrease in grain size from 5-6 μm at 2200°C to 2-3 μm at 2000°C with an increase in hardness.

INTRODUCTION

The ballistic performance of silicon carbide armor ceramics has been shown to be highly dependant on the microstructure. The work of Bakas et al. demonstrated the importance of improving the mixedness of sintering additives in order to eliminate density gradients, defects, and inhomogenieties in the final microstructure[1]. The effect of poorly distributed carbon on exaggerated grain growth has also been shown[2]. The elimination of these negative artifacts typical of silicon carbide green processing is believed to be paramount in improving ballistic performance.

Microstructure modification can also yield positive results. The work of Krell and Straussburger on Al_2O_3 shows a Hall-Petch like relationship displaying increased hardness with decreasing grain size[3]. Work by Cutler et al. demonstrated how modified grain boundary behavior can lead to differing mechanical behavior in similar silicon carbides[4]. Other mechanical properties such as fracture toughness can be modified through advanced processing in order to improve ballistic performance.

Previous work by Ziccardi et al. has addressed the methods and benefits of improved powder processing and its result on microstructure[5]. Cutler et al. suggest aluminum and aluminum containing compounds may strongly influence the mechanical properties of silicon carbide armor ceramics[4]. This method builds upon the previous work to examine if specific aluminum additive groups can be coupled with the boron/carbon surfactant to produce decreased grain sizes and improved properties.

EXPERIMENTAL

A commercially available silicon carbide powder (UK Abrasives) with an average particle size of 1 – 2 μm was used for all experiments. Carbon and boron, as sintering aids, were primarily added through the introduction of a proprietary alkylene amine based surfactant (HX3 – Huntsman Chemical, Austin, Texas) which contains a high percentage of carbon and cross linked boron. Boron carbide (UK Abrasives) and aluminum nitride were added to further alter the sintering behavior and microstructure. Varying amounts of aluminum nitride (Dow Chemical) were added to further alter the sintering behavior and microstructure. The dry powders were mixed and added to an aqueous solution of the surfactant and distilled, deionized water in a polyethylene jar. The slurry was mixed on a ball mill for 3 hours and dried overnight. The powder was sieved and crushed to break up any large agglomerates. Samples were dry pressed into 2.25 inch discs at 35 MPa.

In one set of experiments, samples with a fixed composition of SiC- 2.5 wt% AlN-0.5 wt% B_4C- 3.5% surfactant were loaded into a graphite die and hot pressed over a range of temperatures between 1950°C and 2250°C with a final 15 minute hold at temperature. Samples were held at 1400°C for 90 minutes based upon the research of Kaza, as a sufficient hold at low temperature is required to allow the carbon from the surfactant to reduce the native SiO_2 layer on the SiC[6]. A pressure of 35 MPa was then applied at 1800°C, and held until cool down. In later experiments, the initial composition was adjusted to determine if further improvement in microstructure and properties is possible.

The samples were ground and sectioned, and the density determined using Archimedes principle. Some of the sectioned pieces were polished and analyzed for Knoop hardness. The hardness results are the average of 10 indents taken at each load over a representative section of the sample. Polished samples were etched using a modified Murakami's solution heated to 100°C. Fracture surfaces were prepared on polished samples by indenting with a 10 kg load, and tapping the opposite side with a chisel until fracture occurred. Both fracture and polished samples were mounted and prepared for microscopy. Micrographs were taken using a LEO Gemini 982 FESEM.

RESULTS AND DISCUSSION

Table I shows the composition, density, and hardness for a number of samples. For this composition near theoretical densities were achieved at pressing temperatures of 2100°C and 2200°C, with samples showing hardness values comparable to SiC-N and other similarly processed materials[4]. At the lowest evaluated pressing temperature, 1950°C, the hardness was considerably lower due to the high level of porosity in the sample. At an intermediate temperature of 2000°C, the hardness was higher than expected from the density. This effect could be due to localization of the porosity present in the sample; the variation in hardness reflects the variation in density.

Table 1: Characterization of Materials

Composition	Pressing Temp. (°C)	Density (g/cm3)	HK1 (GPa)
SiC- 2.5 wt% AlN, 0.5 wt% B_4C, 3.5 wt% HX3	2200	3.21	23.4
	2100	3.20	19.6
	2000	3.13	21.6
	1950	2.97	15.6
SiC- 2.5 wt% AlN, 0.2 wt% B_4C, 2.5 wt% HX3	2100	3.2	20.6

A second possibility is that the lower processing temperature has indeed lead to finer grain sizes and increased hardness. Figures 1 and 2 show polished and etched cross sections and fracture surfaces for samples processed at 2200°C and 2000°C respectively. The grain size clearly decreases from 5-6 μm at 2200°C to 2-3 μm at 2000°C. Figure 3 shows surfaces of an AlN-B$_4$C composition taken from Cutler. Differences in fracture behavior can be seen between the larger grained sample, which appears to have a mix of transgranular and intergranular fracture and the finer grained sample, which appears to fracture primarily intergranularly. Both fracture surfaces appear glassier reflecting an increased percentage of intergranular fracture. The fracture surfaces appear different than the surfaces of SiC-N or the compositions explored by Cutler. These differences are likely due to modified grain boundary behavior introduced by the interaction of the aluminum nitride and surfactant.

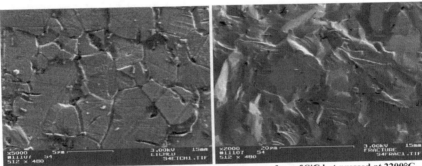

Figure 1: Polished and etched cross section and fracture surface of SiC hot pressed at 2200°C

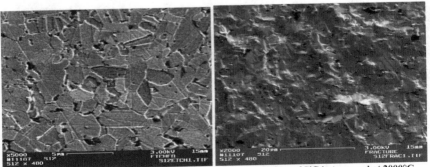

Figure 2: Polished and etched cross section and fracture surface of SiC hot pressed at 2000°C

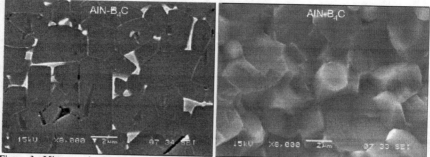

Figure 3: Micrograph of SiC composition hot pressed in work by Cutler[4]

Experiments were also performed on compositions besides those suggested by Cutler[4]. By taking advantage of the combination of the AlN additions and surfactant, new compositions that previously did not lead to high densities were examined. Previous work indicated that additions of less than 3% of the surfactant lead to significantly lower densities in SiC[5]. With a composition of 2.5 wt% AlN, 2.5% surfactant, and 0.2% B_4C, a density of 99.5% was achieved in this work. The micrographs shown in Figure 4 display a polished and etched surface and a fracture surface. Some pores are seen in the material, but they were generally fine and localized. The hardness values reported in Table 1 indicate this composition may also have comparable hardness to SiC-N and other commercial silicon carbide armor ceramics.

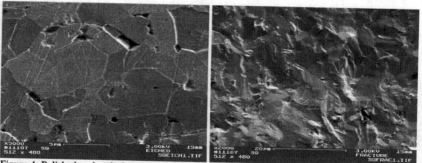

Figure 4: Polished and etched cross section and fracture surface of modified SiC composition hot pressed at 2100°C

CONCLUSIONS

The use of an alkylene amine based surfactant and aluminum nitride as sintering additives for silicon carbide have lead to near theoretical densities at lower processing temperatures. Microscopic evaluation indicates a decrease in grain size from 5-6 μm at 2200°C to 2-3 μm at 2000°C with an increase in hardness. Fracture surfaces indicate a greater percentage of intergranular fracture than materials processed without the AlN and surfactant.

ACKNOWLEDGEMENTS

The authors would like to thank the United States Army Research Laboratory Materials Center of Excellence for the financial support of this work. In addition, thanks go to Joseph Pantina, whose assistance in sample preparation is appreciated.

REFERENCES:

1 M. Bakas, V.A. Greenhut, D.E. Niesz; J. Adams, and J. McCauley. "Relationship between Defects and Dynamic Failure in Silicon Carbide," *Ceramic Engineering and Science Proceedings* 24 no3 351-8 2003.

2 M. Raczka, G. Gorny, L. Stobierski, and K. Rozniatowski. "Effect of Carbon Content on the Microstructure and Properties of Silicon Carbide-based Sinters," *Materials Characterization* 46 (2001) 245-249.

3 A. Krell, "Processing of High-Density Submicrometer Al2O3 for New Applications," *J. Am. Ceram. Soc.*, 86 [4] 546-553 (2003).

4 D. Ray, R. M. Flinders, A. Anderson, and R. A. Cutler. "Effect of Microstructure and Mechanical Properties on the Ballistic Performance of SiC-Based Ceramics," *Advances in Ceramic Armor II, Ceramic Engineering and Science Proceedings, Cocoa Beach, Volume 27, Issue 7* (2006)

5 C. Ziccardi, V. Demirbas, R. Haber, D. Niesz, and J. McCauley "Means of Using Advance Processing to Eliminate Anomalous Defects on SiC Armor", *Ceramic Engineering and Science Proceedings, Vol. 26, No.4,* (2005)

6 A. Kaza. Effect of Gas Phase Composition in Pores during Initial Stages of Sintering. Diss. Rutgers, the State University of New Jersey. (2006).

THE EFFECT OF Si CONTENT ON THE PROPERTIES OF B$_4$C-SiC-Si COMPOSITES

P. Chhillar, M. K. Aghajanian
M Cubed Technologies, Inc.
1 Tralee Industrial Park
Newark, DE 19711

D. D. Marchant
Simula, Inc.
7822 South 46th Street
Phoenix, AZ 85044

R. A. Haber
Center for Ceramic Research
Rutgers University
607 Taylor Road
Piscataway, NJ 08854

M. Sennett
US Army RD&E Command
Natick Soldier RD&E Center
AMSRD-NSC-SS-NS
Kansas Street
Natick, MA 01760

ABSTRACT

Composites of boron carbide, silicon carbide and silicon (B$_4$C-SiC-Si) were fabricated by the reactive infiltration of molten Si into preforms of B$_4$C particles and carbon (the product is often referred to as reaction bonded boron carbide – RBBC). Applications for such materials include armor tiles, neutron absorbing structures, precision components (e.g., where high specific stiffness is required), and wear resistant components. Each of these applications has a unique set of property requirements, with the properties controlled in part by the ratio of each of the constituents in the composite. A major variable is Si content of the final composite. The volume fraction of residual Si is a function of the preform characteristics, namely green density and B$_4$C to carbon ratio. Thus, composites with a wide range of Si contents can be manufactured by altering the preform composition. To assess the effects of Si content on the behavior of the resultant composite, the present work was conducted. B$_4$C-SiC-Si composites with Si contents ranging from nominally 15 to 5 volume percent were fabricated. The composites were characterized for microstructure and composition; and for physical and mechanical properties. The results demonstrate that ceramic-like properties (e.g., hardness, Young's modulus) increase as Si content decreases, whereas fracture toughness drops. Fractography shows a relationship between fracture mode and fracture toughness, with smoother fracture surfaces seen at lower Si contents.

INTRODUCTION

As described elsewhere [1, 2] and shown schematically in Figure 1, reaction bonded B_4C ceramics are produced by the reactive infiltration of molten Si into preforms containing B_4C particles and carbon. During the infiltration process, reaction occurs between the Si and carbon phases, yielding SiC that effectively bonds the B_4C particles into an interconnected ceramic structure. The result is a composite of B_4C, SiC and residual Si. For reference, properties for each of these constituents are provided in Table 1. [Note that when possible, all Knoop hardness measurements were taken with a 2 kg load to provide consistency. This load, however, could not be used with the Si sample, as excessive damage and cracking occurred. Thus, the Si sample was measured with a 200 g load.]

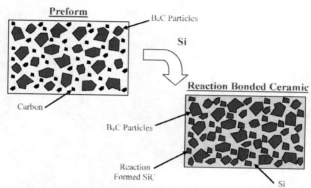

Figure 1. Process Schematic for Reaction Bonded Boron Carbide

Table 1. Properties for Constituents in Reaction Bonded Boron Carbide

	Si	SiC	B₄C
Density (g/cc)	2.33*	3.21**	2.52**
Young's Modulus (GPa)	113*	450**	480**
Knoop Hardness (kg/mm²)	849*** (200 g load)	1905**** (2 kg load)	2069**** (2 kg load)

* Reference [3] – ASM Metals Handbook: Desk Edition
** Reference [4] – ASM Engineered Materials Handbook: Ceramics and Glasses
*** Measured by M Cubed, Polycrystalline Si Sample, Shimadzu HMV-2000 hardness tester
**** Measured by M Cubed, Hot Pressed Samples, Shimadzu HMV-2000 hardness tester

The reaction bonding process offers several advantages relative to traditional ceramic fabrication processes (e.g., hot pressing, sintering). First and foremost, volume change during processing is very low (generally well less than 1%), which provides very good dimensional control and the ability to produce large and complex-shaped components. In addition, the process requires relatively low process temperatures and no applied pressure, which reduces capital and operating costs. Moreover, fine reactive powders capable of being densified are not required, which reduces raw material cost. For these reasons, reaction bonded ceramics are seeing increased use in market areas such as armor, nuclear, wear and precision.

During the molten Si infiltration process, the volume of reaction-formed SiC that is produced is larger than the volume of carbon that is reacted (in the case of graphite to SiC, the volume expansion is 2.3 times). Thus, by infiltrating Si into preforms that contain high carbon contents, ceramic bodies rich in SiC (i.e., low Si content) can be formed.

However, reaction bonded ceramics will always contain some residual Si phase (a requirement of the process is an interconnected Si network to allow the infiltration process to proceed). As clearly seen in Table 1, the presence of Si will result in a decrease in hardness and stiffness of the ceramic composite.

The present work produced reaction bonded B₄C ceramics under multiple processing conditions so as to yield a wide range of compositions (i.e., different final Si contents). The ceramic composites were then characterized to allow the relationship between Si content and properties (microstructure, mechanical properties and fracture behavior) to be quantified. In particular, three formulations were studied, hereinafter referred to as high, mid and low Si content.

EXPERIMENTAL PROCEDURES

All of the reaction bonded ceramics described herein were produced with nominally the same two-step process. First, preforms of B₄C particles and carbon were fabricated by casting a slurry of particles and resin. After casting, the preforms were heated in an inert atmosphere to convert the resin to carbon. An optimized particle size distribution was employed to yield a B₄C content of nominally 75 vol. % in each preform. Second, the resultant preforms were contacted with molten Si alloy in a vacuum atmosphere, thus allowing infiltration to occur. During the infiltration process, the Si reacted with carbon in the preforms to yield SiC. By adjusting the resin type and concentration, preforms with different carbon contents were produced, which after infiltration led to ceramics with different residual Si contents.

After the fabrication step, various mechanical and physical properties of the materials were measured. Density was determined by the water immersion technique in accordance with ASTM Standard B 311. Elastic properties were measured by an ultrasonic pulse echo technique following ASTM Standard D 2845. Hardness was measured on the Knoop scale with a 2 kg load per ASTM Standard C 1236 using a Shimadzu HMV-2000 hardness tester. Flexural strength in four-point bending was determined following ASTM Standard C 1161. Fracture toughness was measured using a four-point-bend-chevron-notch technique and a screw-driven Sintech model CITS-2000 universal testing machine under displacement control at a crosshead speed of 1 mm/min. Specimens measuring 6 x 4.8 x 50 mm were tested with the loading direction parallel to the 6 mm dimension and with inner and outer loading spans of 20 and 40 mm, respectively. The chevron notch, which was cut with a 0.3 mm wide diamond blade, has an included angle of 60° and was located at the midlength of each specimen. The dimensions of the specimen were chosen to minimize analytical differences between two calculation methods according to the analyses of Munz et al. [5].

Microstructure was characterized in two manners. Polished sections were examined using a Nikon Microphot-FX optical microscope and fracture surfaces were studied with a JOEL 840 scanning electron microscope (SEM). Finally, phase composition was characterized by quantitative image analysis (QIA) and quantitative x-ray diffraction (XRD). The QIA was performed using a Clemex JS-2000 digital scanning system, and XRD was conducted with a Phillips PW3020 diffractometer.

RESULTS AND DISCUSSION
Microstructure and Composition
Optical photomicrographs of three reaction bonded B_4C samples are provided in Figure 2. In each case, the particulate phase is B_4C, the bright matrix phase is Si, and the dark matrix phase is SiC. A slight halo exists around many of the B_4C particles as a result of Si-B_4C reaction. In each of the ceramic composite microstructures, the B_4C particle size distribution (median particle size of nominally 50 microns with a range of about 10 to 100 microns) and the B_4C content are about the same. However, there is a significant difference in Si content.

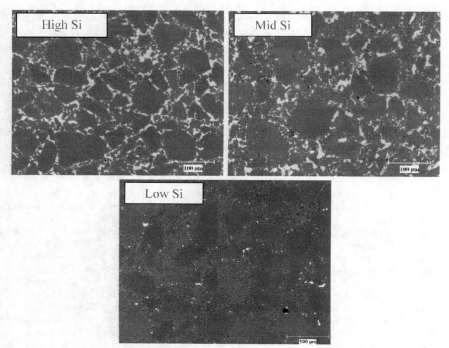

Figure 2. Optical Photomicrographs of Reaction Bonded B_4C Ceramic Composites with Differing Si Contents

Compositional data are provided in Table 2. Good agreement with respect to Si content is seen between the two analysis methods – QIA and XRD. Moreover, the ability to greatly adjust the Si content by process modification is demonstrated with nearly a 3 times difference between the high and low Si content composites. The XRD result appears to slightly over estimate the B_4C to SiC ratio in the composites. Based on green density of the preforms, the B_4C content should be in the range of 72 to 75 vol. %, not the approximately 80 vol. % as predicted by XRD.

Table 2. Compositional Data for Three Composites Shown Microstructually in Figure 1

Test	High Si Content	Mid Si Content	Low Si Content
Quantitative Image Analysis (results in vol. %)	Si Content: 14 % Ceramic Content: 86 %	Si Content: 10 % Ceramic Content: 90 %	Si Content: 5 % Ceramic Content: 95 %
Quantitative XRD (results in wt. %)	Silicon: 13.3 % Silicon Carbide: 8.7 % Boron Carbide: 78.0 %	Silicon: 8.2 % Silicon Carbide: 14.3 % Boron Carbide: 77.5 %	Silicon: 4.6 % Silicon Carbide: 16.7 % Boron Carbide: 76.5 %
Quantitative XRD (results converted to vol. %)	Silicon: 14.5 % Silicon Carbide: 6.9 % Boron Carbide: 78.6 %	Silicon: 9.1 % Silicon Carbide: 11.5 % Boron Carbide: 79.4 %	Silicon: 5.1 % Silicon Carbide: 13.0 % Boron Carbide: 81.9 %

Mechanical and Physical Properties

Results of density, Young's modulus, hardness, flexural strength and fracture toughness measurements are provided in Table 3. When appropriate, the results are provided as a mean ± one standard deviation. The density of the composites increases as the Si content decreases as Si is the lowest density phase present. Similarly, Young's modulus and hardness increase as Si content decreases, which follows the data in Table 1 showing the low properties for Si relative to SiC and B_4C. The hardness of the low Si composite exceeds that of hot pressed SiC and approaches that of hot pressed B_4C (compare data in Tables 1 and 3).

Table 3. Property Data for Three Composites Shown Microstructually in Figure 1

Property	High Si Content	Mid Si Content	Low Si Content
Density (g/cc)	2.56	2.58	2.62
Young's Modulus (GPa)	380	406	432
Knoop 2 kg Hardness (kg/mm2)	1556 ± 113	1704 ± 158	1992 ± 45
Flexural Strength (MPa)	276 ± 22	280 ± 15	324 ± 14
Fracture Toughness (MPa-m$^{1/2}$)	5.0 ± 0.6	4.8 ± 0.5	4.4 ± 0.3

Previous work [1] has shown high fracture toughness of reaction bonded B_4C ceramics, with the high toughness attributed to ductile-like failure of the Si phase. The same response is seen in the present work with all reaction bonded B_4C ceramics showing fracture toughness values greater than 4 MPa-m$^{1/2}$, compared to values of 2.9 to 3.7 MPa-m$^{1/2}$ for hot pressed B_4C [6]. A trend of decreasing toughness with decreasing Si content is seen, although the result is not statistically significant. Finally, strength of the low Si content composite is higher than that of the higher Si content composites, despite lower fracture toughness. This suggests reduced flaw size in this material.

Analysis of Fracture Surfaces

SEM fractographs of the high, mid and low Si content composites are provided in Figure 3. Significant difference between the three fracture surfaces are seen. The fracture surface of the high Si content composite shows two significant features. First, the fracture is very non-planar, and second, a high level of deformation of the Si second phase is seen. Both of these fracture characteristics lead to high toughness, which is seen with this formulation.

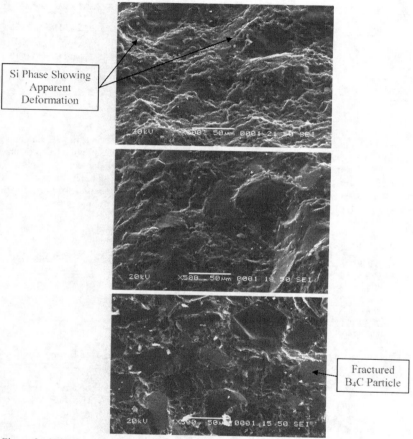

Si Phase Showing Apparent Deformation

Fractured B₄C Particle

Figure 3. SEM Fractographs of Reaction Bonded B₄C Ceramics include High Si Content (top); Mid Si Content (middle); and Low Si Content (bottom)

The fracture surface of the low Si content composite shows a relatively planar crack path and little deformation of the second phase. In this fracture surface, the B₄C particles, which failed transgranularly along the plane of the propagating crack, are clearly visible. In the high Si content composite, it is far more difficult to observe the B$_4$C particles as they are obscured by deformation of the Si second phase. The mid Si content composite shows intermediate fracture behavior.

The fracture behavior of the composites is consistent with the measured properties shown in Table 3. The trend of decreasing toughness with decreasing Si content can be explained due to the less tortuous fracture path (more planar) and less second phase deformation in the low Si content ceramic. Moreover, the high strength of the low Si composite can be explained by the apparent smaller critical flaw size in this material. That is, fracture surface features in the low Si content material are on the scale of the B$_4$C particles, whereas the fracture surface features on in the higher Si content materials are significantly larger – on the scale of multiple particles.

SUMMARY

Due to processing advantages relative to hot pressed B$_4$C, reaction bonded B$_4$C ceramics are seeing increased use in multiple markets, including armor plates, neutron absorbing structures, wear components and precision equipment. However, the presence of a relatively soft and low stiffness Si second phase hurts some mechanical and physical properties, thus limiting utility. The present work examined new, lower Si content versions of reaction bonded B$_4$C. Reduction in Si content led to significant increases in hardness and Young's modulus, nearly matching results seen with hot pressed B$_4$C. As expected, fracture toughness dropped as Si content was reduced. Nonetheless, the toughness of all the reaction bonded ceramics exceed that of hot pressed B$_4$C. Moreover, the scale of fracture surface features was reduced via a reduction in Si content, which led to an increase in strength. In summary, the properties of reaction bonded ceramics differ from those of traditional hot pressed and sintered ceramics due to the presence of a Si second phase. By reducing Si content, the resultant reaction bonded ceramic displays properties that approximate those of a traditional ceramic.

ACKNOWLEDGEMENTS

This work was partially supported by Army/Natick Labs under contract number W911QY-06-C-0041.

REFERENCES

1. M. K. Aghajanian, B. N. Morgan, J. R. Singh, J. Mears, R. A. Wolffe, "A New Family of Reaction Bonded Ceramics for Armor Applications", in Ceramic Armor Materials by Design, *Ceramic Transactions*, **134**, J. W. McCauley et al. editors, 527-40 (2002).
2. P. G. Karandikar, M. K. Aghajanian and B. N. Morgan, "Complex, Net-Shape Ceramic Composite Components for Structural, Lithography, Mirror and Armor Applications, *Ceram. Eng. Sci. Proc.*, **24** [4] 561-6 (2003).
3. *Metals Handbook: Desk Addition* (ASM International, Metals Park, OH, 1985).
4. *Engineered Materials Handbook, Vol. 4. Ceramics and Glasses*, (ASM International, Metals Park, OH, 1991).
5. D. G. Munz, J. L. Shannon and R. T. Bubsey, "Fracture Toughness Calculations from Maximum Load in Four Point Bend Tests of Chevron Notch Specimens", *Int. J. Fracture*, **16**, R137-41 (1980).
6. F. Thevenot, "Boron Carbide – A Comprehensive Review", *Euro-Ceram, Vol. 2, Properties of Ceramics*, G. de With, R. A. Terpstra and R. Metselaar, editors, 2.1-2.25 (Elsevier Applied Science, London and New York, 1989).

Damage and Testing

PRELIMINARY INVESTIGATION OF DAMAGE IN AN ARMOR-GRADE B$_4$C INDUCED BY QUASI-STATIC HERTZIAN INDENTATION

R.C. McCuiston[*], H.T. Miller[**], and J.C. LaSalvia

U.S. Army Research Laboratory
AMSRD-ARL-WM-MD
Aberdeen Proving Ground, MD 21005-5069

ABSTRACT

Previous work on sphere impact induced damage in a commercially-available armor-grade boron carbide (B$_4$C) revealed that the mechanisms (e.g. shear localization) that govern compressive inelastic response are significantly different than either hot-pressed polycrystalline SiC (grain boundary microcracking) or WC (plasticity). To further probe the inherent damage mechanisms present in polycrystalline B$_4$C, quasi-static Hertzian indentation was used to generate a series of controlled indentations as a function of load. B$_4$C specimens were sectioned from commercially-available armor-grade material and final polished using colloidal silica. The specimens were then subjected to a series of Hertzian indentations using a Zwick/Roell Z005 MTM with spherical diamond indenters with radii of 1.5 and 2.5 mm. Indents were made at loads between 500 N and 2000 N at a constant crosshead speed of 1.0 μm/sec. Optical and electron microscopy was performed on the indented surfaces and on cross-sections taken through the indents. Results on the evolution of damage versus the applied load will be reported and discussed.

INTRODUCTION

The application of B$_4$C armor ceramic is currently limited to small caliber threats, due in part to its extremely brittle nature and sub-optimal ballistic efficiency, when compared to other armor ceramics for certain threats. As the U.S. Army seeks to decrease the weight of future combat vehicles, the use of monolithic or composites based on B$_4$C has become more important. The U.S. Army Research Laboratory (ARL) is currently working[1] in conjunction with other laboratories to understand the root material causes for B$_4$C's ballistic behavior, with the long term goal of improved B$_4$C based armor ceramics.

It has long been known from plate impact experiments that B$_4$C suffers an apparent loss of shear strength in the range of 20-23 GPa, but the reason(s) for the loss remained unexplained.[2-3] Recently the existence of a pressure induced localized amorphization in B$_4$C has been reported.[4-6] A density function theory (DFT) simulation of B$_4$C under stress has identified a B12-CCC polytype of B$_4$C that collapses under a moderately low pressure, leading to localized amorphization.[7] The pressure induced localized amorphization is postulated to be a mechanism responsible for B$_4$C's loss of shear strength, but it is not yet conclusive. More recently, a shear localization phenomenon has been identified in B$_4$C that was subjected to sphere impact, which manifests itself as shear bands containing a nanoscale comminuted B$_4$C.[8] Raman spectroscopy

[*,**] Work performed with support by an appointment to the Research Participation Program at the U.S. ARL administered by the [*]Oak Ridge Associated Universities and the [**]Oak Ridge Institute for Science and Education through an interagency agreement between the U.S. Department of Energy and U.S. ARL.

conducted on and around the shear bands showed evidence of amorphization which may be spatially associated with the pressure induced localized amorphization mechanism.

In an effort to better understand and quantify the newly discovered damage generation mechanisms in B₄C, the U.S. ARL has begun Hertzian indentation studies to compliment sphere impact studies all ready underway.[9-11] A Hertzian indentation made using a spherical indenter tests both the elastic and plastic response of the material, rather than just the plastic response as with a Vicker's or Knoop indent.[12] A typical ballistic event, while transient in nature, does progress from an elastic response to a plastic response. The material far from the impact will also experience an elastic response, from the propagating stress waves, so it is important to have the ability to also test the elastic response. Two benefits of the spherical Hertzian indentation technique, as well as others indentation techniques in general, over sphere impact, is that it can be readily instrumented[13-15] and the sample is easily recovered for further evaluation. Another benefit unique to elastic Hertzian indention, as it does not involve large strain rates and transient wave generation like the sphere impact, is that the stress state under the indenter can be readily described using the Hertz equations[17] for contact, several of which are shown in Table I.

Table I. Partial list of Hertzian contact equations for a circular contact area.[17]

Contact Modulus E* (GPa)	Effective Curvature R (m)
$E^* \equiv \left(\dfrac{1-v_1^{\,2}}{E_1} + \dfrac{1-v_2^{\,2}}{E_2} \right)^{-1}$	$R \equiv \left(\dfrac{1}{R_1} + \dfrac{1}{R_2} \right)^{-1}$
Contact Radius a (m)	Mean and Max Contact Pressure (GPa)
$a = \left(\dfrac{3PR}{4E^*} \right)^{1/3}$	$P_{max} = \left(\dfrac{3P}{2\pi a^2} \right) = \left(\dfrac{6PE^{*2}}{\pi^3 R^2} \right)^{1/3} \quad P_{mean} = \dfrac{2}{3} P_{max}$
Maximum Tensile Stress σ_{max} (GPa)	Maximum Shear Stress τ_{max} (GPa)
$\sigma_{max} = \dfrac{1}{2}(1-2v_1)P_{mean}$	$\tau_{max} = 0.48 P_{mean}$
$v_{1,2}$ = Poisson's Ratio $\quad\quad$ $R_{1,2}$ = Radius (m) $E_{1,2}$ = Young's Modulus (GPa) $\quad\quad$ P = Indenter Load (N) The subscripts 1 and 2 represent the sample and the indenter material respectively	

A cross section of a B₄C cylinder tested using the sphere impact technique is shown in Figure 1 (a). A cone crack is readily visible, as are several ring cracks adjacent to it on both sides. There is also significant erosion around the tip of the cone. Figure 1 (b) shows a diagram of the various forms of damage induced in B₄C using a spherical Hertzian indenter, that were found in this study. The top surface of an indentation typically shows the primary central ring crack, followed by a series of concentric ring cracks. Between the ring cracks are several inter-ring radial cracks and spalls. The cross section of the indentation shows the main cone crack, the conoid and several ring cracks. In several instances, sub-surface cone cracks were formed below the main cone crack. The formation of a cone crack and concentric ring cracks, during spherical

Hertzian indentation, are the hallmarks of an extremely brittle material like B_4C or glass.[15,16] The similarity between the damage generated by the sphere impact test and the spherical Hertzian indentation is apparent and thus the compliment of spherical Hertzian indentation is our reason for pursuing it.

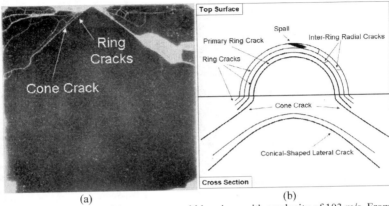

(a)   (b)

Figure 1. (a) B_4C impacted with a tungsten carbide sphere with a velocity of 103 m/s. From Ref. 10. The impact generated sub-surface damage is very similar to a controlled spherical Hertzian indentation. (b) Diagram of the representative top and sub-surface damage generated in B_4C in this study.

EXPERIMENTAL PROCEDURE

Two samples for indentation were sectioned from a commercially available, hot pressed B_4C tile (PAD B_4C, BAE Systems, Vista CA) into blocks roughly 25 x 25 x 12 mm. The two samples were mounted in thermoplastic resin (IsoFast, Struers, Cleveland OH). The mounted samples were polished using standard metallographic techniques to a finish of 1 μm. The polished faces of the samples were oriented such that they were parallel to the hot pressing direction.

The polished samples were subjected to a series of Hertzian indentations using spherical diamond indenters of radii 1.5 and 2.5 mm. A universal test machine (Z005, Zwick/Roell, Atlanta GA) with a 2500 N load cell was used to generate the indents. The cross head speed was a constant 1 μm/sec for loading and unloading. Four loads, 500, 1000, 1500 and 2000 N were used.

The indentations, for both the top surface and cross sections, were examined optically and using a scanning electron microscope (S4700, Hitachi, Pleasanton CA). The cross sections were prepared by sectioning the samples several millimeters from the center of the indentations using a diamond wafering blade. The samples were then polished to the center of the indentation using standard metallographic techniques.

RESULTS AND DISCUSSION

Assuming values of Young's modulus and Poisson's ratio for B$_4$C and diamond as E_1 = 450 GPa, v_1 = 0.16 and E_2 = 1141 GPa, v_2 = 0.07 respectively, the mean and maximum elastic contact pressures were calculated from the equations in Table I and are shown in Table II. These values are only valid until the yield point of the B$_4$C is reached and are provided for reference only. However, the values of the contact pressure can be substantial and are well within the range of a sphere impact.

Table II. The calculated elastic mean and max contact pressures in B$_4$C under the 1.5 and 2.5 mm radii diamond indenters.

Load →	500 N		1000 N		1500 N		2000 N	
Indenter Radius	Mean (GPa)	Max (GPa)	Mean (GPa)	Max (GPa)	Mean (GPa)	Max (GPa)	Mean (GPa)	Max (GPa)
1.5 mm	11.1	16.7	14.0	21.0	16.1	24.1	17.7	26.5
2.5 mm	7.9	11.9	10.0	15.0	11.4	17.1	12.6	18.9

Optical images of the top surface of the four indentations, made using the 2.5 mm radius indenter, are shown in Figure 2 (a). The black inclusions visible in the four images, as well as the remaining images that will be presented, are graphitic carbon and/or sintering additive inclusions. The image of the 500 N indentation shows a barely visible primary ring crack, surrounded by even fainter concentric ring cracks. It is common in indents made below a certain load for the primary ring crack and concentric rings cracks to close almost perfectly, making them hard to observe. Increasing the load to 1000 N makes the primary ring crack and concentric rings cracks more visible, as they have closed imperfectly, leaving visible gaps. The "roundness" of the ring cracks is far from perfect. The crack paths, at this magnification, appear to be influenced by the graphitic carbon and sintering additive inclusions, thus affecting the "roundness". The "roundness" may also be influenced by the elastic anisotropy of the diamond indenter.[18] When the load is increased to 1500 and then 2000 N, there is an increase in the number of concentric ring cracks. The indent at 2000 N also shows the generation of inter-ring radial cracks, diagramed in Figure 1 (b). In the lower right quadrant of the 2000 N indent, a spall is visible, but it has not been removed. The creation of multiple concentric ring cracks is the hallmark of a brittle material, such as glass, and points to the brittle nature of B$_4$C. By comparison, a spherical indentation placed in a more "quasi-plastic" ceramic, such as silicon nitride, will result in a smooth, hemispherical indent, with substantially less cracking. At a microstructural level, this indentation behavior is due to intra-granular cracking and grain boundary sliding, both of which act as quasi-plasticity mechanisms.

The cross sections, showing the sub-surface damage induced in the samples using the 2.5 mm radii indenter, are show in Figure 2 (b). Lines have been drawn on either side of the cone crack, in the optical image of the 500 N indent, to show its location. As also seen on the top surface image, the cone crack in the 500 N indent has closed almost perfectly, making it hard to observe. An increase of the load to 1000 N, results in a very visible cone crack, as well as several ring cracks. The indent at 1500 N shows the main cone crack and several ring cracks, but there is also a very prominent cone-shaped or conical lateral crack, as diagramed in Figure 1 (b). The conical lateral crack may form upon unloading, due to a slip-stick behavior. Upon unloading, the conoid, i.e. the cone of material below the cone crack, moves upward due to a release of stored elastic energy, closing the cone crack. The upward motion of the conoid may be temporarily halted, due to frictional forces generated on a microstructural level by displaced grains, among other mechanisms. Upon continued removal of the load, the conoid may suddenly slip upwards, generating tensile forces within, causing the formation of the conical lateral crack. The generation of conical lateral cracks requires further study to determine their exact formation mechanism. The final image in the series, of the 2000 N indent, shows the cone crack and multiple ring cracks. No conical lateral crack was generated. It should be noted that no sub-surface damage associated with quasi-plasticity can be observed. This again points to the extremely brittle nature of B$_4$C.

The top surface optical images of the indents generated using the 1.5 mm radius indenter are shown in the Figure 3 (a). The smaller radius of the 1.5 mm indenter generates larger contact pressures, which should manifest as increased damage. The primary ring crack and the concentric ring cracks in the 500 N indent, qualitatively, are slightly more visible than the 500 N indent made with the 2.5 mm radius indenter. The indent made at 1000 N shows very visible ring cracks. Again, the "roundness" of the rings crack appears to be influenced by the graphitic carbon and sintering additive inclusions. There is a visible spall that has been removed from the lower right quadrant on the indent. The indent made at 1500 N shows an increased number of concentric ring cracks and visible inter-ring radial cracks. There are also two removed spalls. The indent at 2000 N, made with the 1.5 mm radius indenter, should show the most damage, as it experienced the highest contact pressure. Indeed, there are a large number of concentric ring cracks and several inter-ring radial cracks. There are also three small spalls visible. The increase in load from 500 to 2000 N has generated an increase in the number of ring and inter-ring radials cracks, as well as the number of observed spalls.

Optical images of the cross sections for the indents, generated using the 1.5 mm radius indenter, are shown in Figure 3 (b). The cross section of the 500 N indent very clearly shows the cone crack and several of the concentric ring cracks. The cross section of the 1000 N indent does not clearly show the cone crack and the ring cracks and is likely due to over zealous final polishing of the sample. This also occurred in the 1500 and 2000 N indents, unfortunately. The indent generated at 1500 N shows the expected cone crack and several concentric ring cracks. There is also a noticeable conical lateral crack with another less noticeable conical lateral crack directly over it. The presence of two conical lateral cracks could again be explained by the previously described hypothesis of a slip-stick event generating tensile forces. The indent generated at 2000 N does contain a cone crack and several ring cracks, though they are hard to observe in the image. There are also several triangular sections removed from the top surface, likely during polishing. The ease of removal, though, points to the extensive cracking that is present. The 2000 N indent exhibits a noticeable conical lateral crack. There are also two less noticeable conical lateral cracks above it that are shorter in length. Within the noticeably large conical lateral crack appear two regions of comminuted material, a short distance away from the center axis of the indent. The generation of the comminuted zones is of interest for further study. As with the previous series of cross sections using the 2.5 mm radius indenter, there is no noticeable sub-surface damage associated with quasi-plasticity mechanisms.

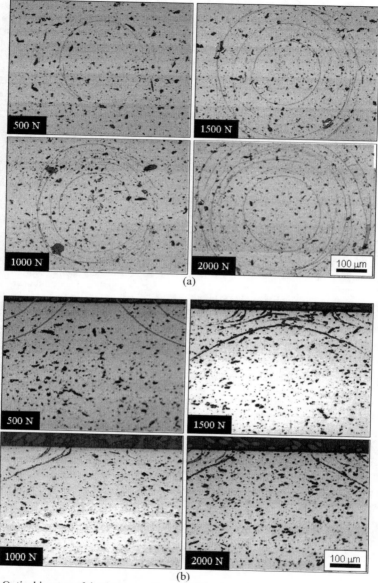

Figure 2. Optical images of the (a) top surface and (b) sub-surface of the indents using the 2.5 mm radius indenter for the indicated loads.

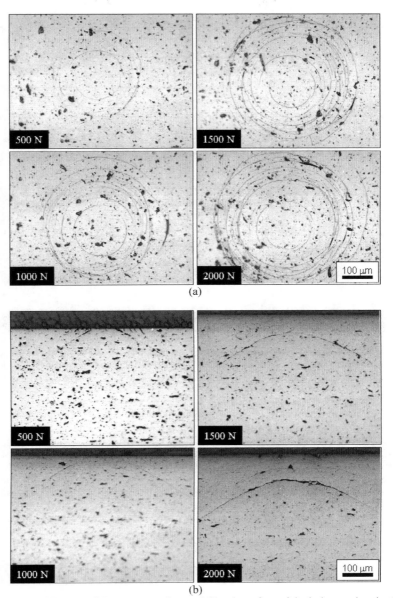

Figure 3. Optical images of the (a) top surface and (b) sub-surface of the indents using the 1.5 mm radius indenter for the indicated loads.

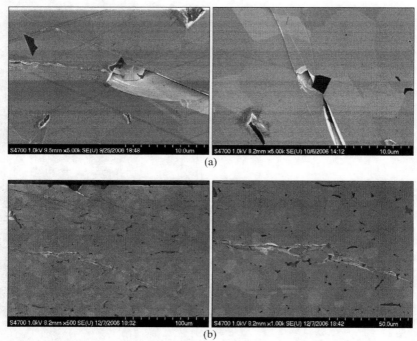

(a)

(b)

Figure 4. (a) SEM images showing the origin of two representatives spalls in B₄C indents. The spalls originate from graphitic and sintering additives inclusions. (b) SEM images showing the sub-surface damage in the 1.5 mm radius 2000 N indent showing the conical lateral crack containing a small region of comminuted B₄C.

Figure 4 (a) shows electron microscope images of two top surface spall sites, from the indent generated at 2000 N using the 1.5 mm radius indenter, as seen in Figure 3 (a). The image on the left of Figure 4 (a) shows an inter-ring radial crack that has been generated at a graphitic carbon inclusion. The inter-ring crack allowed for the removal of a spall. There is also what appears to be cleavage steps associated with the trans-granular failure of a B₄C grain. The image on the right of Figure 4 (b) shows a small spall from the corner of a sintering additive inclusion. It is also apparent that the ring crack nearby is passing through the B₄C grains in a trans-granular fashion.

Figure 4 (b) shows electron microscope images of the cross section of the indent generated at 2000 N, using the 1.5 mm radius indenter, as seen in Figure 3 (b). The image on the left of Figure 4 (b) shows the right half of the main cone crack, several of the ring cracks and the large conical lateral crack that was generated. It is obvious that there is no heavily micro-cracked region associated with quasi-plasticity. The image on the right of Figure 4 (b) shows the small comminuted region that was generated within the conical lateral crack. The comminuted material is on the order of the grain size and is comprised of 5 or 6 grains. Along with the generation of the conical lateral crack, the generation of this small comminuted region warrants further study.

SUMMARY AND CONCLUSIONS

Visual observations of spherical Hertzian indentations in B$_4$C have shown that B$_4$C has a brittle response. Images taken of the top surface of the indents show a primary ring crack and several concentric ring cracks, damage consistent with a brittle material. The number of concentric ring cracks and inter-ring radial cracks increased with indentation load, as did the number of observed spalls. Images of the cross sections of the indents also showed well formed cone crack and ring cracks. There was no observed damage associated with quasi-plasticity. In several of the indents, conical lateral cracks were formed, which may be the results of a slip-stick behavior. In the cross section of the 2000 N indent formed using the 1.5 mm radius indenter, a small region of comminuted material, on the order of the grain size, was formed within a conical lateral crack. The generation of the conical lateral cracks and the small comminuted region may be of importance to damage evolution in B$_4$C and warrants further study.

ACKNOWLEDGEMENT

The authors would like to thank Mr. Matthew J. Motyka (Dynamic Science International) for his technical support.

REFERENCES

1. T.J. Moynihan, J.C. LaSalvia and M.S. Burkins, "Analysis of Shatter Gap Phenomenon in a Boron Carbide/Composite Laminate Armor System", In *20th Int. Symposium on Ballistics*, Vol II., eds. J. Carleone and D. Orphal, 1096-1103 (2002)
2. M.E. Kipp and D.E. Grady, "Shock Compression and Release in High-Strength Ceramics", Sandia Report SAND89-1461, (1989)
3. D.E. Grady, "Dynamic Properties of Ceramic Materials", Sandia Report SAND94-3266, (1994)
4. M. Chen, J.W. McCauley and K.J. Hemker, "Shock-Induced Localized Amorphization in Boron Carbide", *Science*, **299**[5612], 1563-66, (2003)
5. V. Domnich, Y. Gogotsi, M. Trenary and T. Tanaka, "Nanoindentation and Raman Spectroscopy Studies of Boron Carbide Single Crystals", *Appl. Phys Letters*, **81**[20] 3783-85 (2002)
6. D. Ge, V. Domnich, T. Juliano, E.A. Stach and Y. Gogotsi, "Structural Damage in Boron Carbide Under Contact Loading", *Acta Mat.*, **52** 3921-27 (2004)
7. G. Fanchini, J.W. McCauley and M. Chhowalla, "Behavior of Disordered Boron Carbide under Stress", *Phys. Rev. Letters*, **97** 035502 (2006)
8. J.C. LaSalvia, R.C. McCuiston, G. Fanchini, M. Chhowalla, H.T. Miller, D.E. MacKenzie and J.W. McCauley, "Shear Localization in a Sphere-Impacted Armor Grade Boron Carbide", To be published: *Proc. of the 23rd Int. Symposium on Ballistics*, Tarragona, Spain, (2007)
9. J.C. LaSalvia, M.J. Normandia, H.T. Miller and D.E. MacKenzie, "Sphere Impact Induced Damage in Ceramics: I. Armor Grade SiC and TiB$_2$", *Cer. Eng. Sci. Proc.*, **26**(7), 171-79 (2005)
10. J.C. LaSalvia, M.J. Normandia, H.T. Miller and D.E. MacKenzie, "Sphere Impact Induced Damage in Ceramics: II. Armor-Grade B$_4$C and WC", *Cer. Eng. Sci. Proc.*, **26**(7), 180-88 (2005)
11. J.C. LaSalvia, M.J. Normandia, D.E. MacKenzie and H.T. Miller, "Sphere Impact Induced Damage in Ceramics: III. Analysis", *Cer. Eng. Sci. Proc.*, **26**(7), 189-98 (2005)
12. B.R. Lawn, "Indentation of Ceramics with Spheres: A Century after Hertz", *J. Am. Ceram. Soc.*, **81**(8), 1977-94 (1998)
13. A.A. Wereszczak and R.H. Kraft, "Instrumented Hertzian Indentation of Armor Ceramics", *Cer. Eng. Sci. Proc.*, **23** 53-64 (2002)

14. A.A. Wereszczak and K.E. Johanns, "Spherical Indentation of SiC", *Cer. Eng. Sci. Proc.*, **27**(7), (2006)

15. R.F. Cook and G.M. Pharr, "Direct Observation and Analysis of Indentation Cracking in Glasses and Ceramics", *J. Am. Ceram. Soc.*, **73**(4), 787-817 (1990)

16. A.A. Wereszczak, K.E. Johanns, T.P. Kirkland, C.E. Anderson, T. Behner, P. Patel and D.W. Templeton, "Strength and Contact Damage Responses in a Soda-Lime-Silicate and a Borosilicate Glass", *Proc. of the 25th Army Sci. Conf.*, Orlando, FL, (2006)

17. K.L. Johnson, Contact Mechanics, Cambridge University Press, New York, (1985)

18. Personal communications with A.A. Wereszczak.

IN-SITU FRAGMENT ANALYSIS WITH X-RAY COMPUTED TOMOGRAPHY, XCT

J.M. Wells, Sc.D.
Principal Consultant, JMW Associates, 102 Pine Hill Blvd., Mashpee, MA 02649
(774)-836-0904   jmwconsult1@comcast.net

ABSTRACT
    Fragmentation characterization is of importance to blast and ballistic protection analysis. In the case of the fragmentation of the active element – an explosive device or the high velocity ballistic projectile disintegrating on impact – the fragments may be of heavy metallic alloy remnants imbedded and distributed internally within the target material. Alternatively, one may be interested in the fragmentation of the impacted target material itself, particularly if composed of a brittle armor ceramic material. The in-situ high resolution, volumetric, and quantitative diagnosis of residual fragments is a difficult technical challenge unsatisfactorily accomplished with destructive sectioning or conventional NDE modalities. Recent investigations using x-ray computed tomography, XCT, diagnostic interrogation of residual fragments in various impacted armor ceramics have demonstrated revealing qualitative and quantitative results on the location, size, morphology, metrology, and structural complexity of embedded high K.E. projectile fragments. Impacted ceramic targets are x-ray scanned to generate volumetric digitization files that are subsequently imported as 16-bit TIFF or JPEG formatted images into, and analyzed with, Volume Graphics StudioMax v. 1.2.1 voxel analysis and visualization software. Results to date of these evolving XCT diagnostic fragmentation analysis techniques are summarized and presented herein along with some recommendations for further investigative efforts.

INTRODUCTION

Ceramic armor is traditionally designed to operate on the edge of failure – it is made as light as possible and provides the required threat-specific level of protection, but no more! The two key consequences of ballistic impact experienced by the armor ceramic are penetration and/or structural damage. The consequence of penetration is unforgiving as you either stop the projectile or you don't. The inevitable consequence of damage alone is more forgiving as the initial damage level may not be catastrophic on the first hit and sufficient structural integrity remains to endure further impacts. High kinetic energy, K.E., impacting projectiles and the impacted armor ceramic targets experience various degradation processes during the impact event. One common degradation event often experienced by both the projectile and the target ceramic is that of fragmentation.

Fragmentation is the process of breaking up a larger size solid object into multiple smaller sized segments of that same object. Among the separate and distinct fragmentation processes occurring during terminal ballistic events on armor ceramic targets are included:

- Fragmentation of the impacting projectile, either externally (as dwell with zero penetration or behind armor debris with full penetration) or internally (as embedded projectile remnants) with respect to the target ceramic.
- Fragmentation of the interior structure of the host ceramic target material, either as meso-scale fragments and/or as micro-scale comminution.
- Fragmentation and ejecta from the rear surface of the target material (spall).

Ideally, one desires for the projectile to dwell and completely self-destruct on the impact surface of the target ceramic without the occurrence of any significant penetration - a condition referred to as "interface defeat". Unfortunately, this condition is seldom realized except under rigorous laboratory testing conditions. Thus, the more common practical concern is to limit the extent of penetration to that substantially less than the full thickness of the armor material. Impact-induced internal ceramic damage: occurs prior to any projectile penetration, is essentially controlled by the damage tolerance attributes of the target ceramic, and may have a direct influence on the initiation and progression of penetration [1]. Much research has been focused upon the extent and the rate of penetration behavior with technical ceramics under various ballistic conditions. However, penetration observations provide little additional information that is directly useable for deciphering intrinsic armor ceramic damage mechanisms, or suggestive of corrective material or design architectural modifications. Thus if one has to accept the inevitable occurrence of some penetration, one needs an alternative approach of establishing cognizant methods for its mitigation. One such approach is that of ballistic impact damage analysis.

Impact damage is defined here as diverse irreversible physical changes of the target material, structural continuity, and/or structural integrity resulting from a specific impact event. Among the several physical changes frequently observed within an impacted armor ceramic target is the presence of residual projectile fragments. These fragments are normally observed post-impact with the occurrence of either partial or full penetration, and are remnants of the high density K.E. projectile material remaining embedded within the target ceramic. Most often such residual fragments are not easily detectable without physical sectioning or interrogation by nondestructive examination, NDE, methods. Even when detected, seldom have the detailed quantities, size, shape, morphology, and spatial distribution characteristics of the residual projectile fragments been reported. Potential consequential effects of embedded fragments may include:

-   damaged and displaced host ceramic material
-   mixing and reconsolidation of fine ceramic debris with fragment debris
-   wedging – preventing localized ceramic relaxation
-   evidence of fragment flow into ceramic fissures
-   impact-induced porosity found in near surroundings of embedded fragments

The presence of residual projectile fragments embedded within the target ceramic interior may be considered as an indication of the effectiveness of that ceramic to stop the full penetration of the entire projectile. The aim of this presentation is to demonstrate both qualitative and quantitative representative results obtained in the characterization of embedded high density tungsten alloy fragments within various armor ceramic targets using X-ray computed tomography, XCT, diagnostic techniques. Details of the fragmentation of the host ceramic are not discussed in this paper.

XCT RESIDUAL PROJECTILE FRAGMENT DIAGNOSTICS

Three impacted ceramic targets from the earlier ballistic studies of D.A. Shockey et al. [2-4] were made available and recently re-examined with XCT. The original ceramic target samples were longitudinally sectioned and a half section of each was interrogated for the present study. The sample dimensions and visual observations of fragment penetration are listed in Table 1 for each of the three sample ceramic targets.

Table I. Measured Dimensions & Projectile Penetration Observations

Ceramic Target	Height mm	Diameter (Width) mm	Projectile Penetration Observations
B₄C	112.8 – 114.2	102.9 –104.2	Full & Intermittent
Al₂O₃	99.9 – 100.1	102.4 – 102.8	Full & Discontinuous
TiB₂	90.2 – 92.7	98.7 – 101.9	Partial & Intermittent

Following digitization via a series of non-invasive XCT scans, the subsequent virtual characterization, analyses, and visualizations of selected impact damage features are conducted using an assortment of image processing and analysis tools included in the Volume Graphics StudioMax v 1.2.1 software package [5]. This software provides powerful segmentation (virtual partitioning a volume data set into separate regions) capabilities to isolate and characterize internal structures or features by both grey level values and/or by morphological characteristics. The application of various XCT diagnostic techniques has been demonstrated to permit the noninvasive detection, location, segmentation, in-situ characterization, metrology and the 3D visualization of multiple impact damage features including embedded projectile fragments [6-10].

BORON CARBIDE TARGET

Since the K.E. projectile fragment material is of considerably higher density (higher grey level) than either the host ceramic or the other damage manifestations, the fragment remnants were initially isolated directly with a high grey level segmentation operation. This is the simplest of the segmentation processes available within the advanced voxel analysis and visualization software [5] utilized for these analyses. Figure 1 shows a reconstructed opaque virtual 3D solid object image of the B₄C half-cylinder ceramic target and then a virtually transparent overview of the embedded and segmented projectile fragments within a segmented hemi-cylindrical epoxy outer coating.

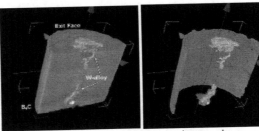

Figure 1. Reconstructed virtual XCT images of B₄C ceramic target in an opaque view (left) and a transparent view (right) clearly reveal the embedded projectile fragments with an apparent absence of fragments at the target mid-thickness.

It is readily apparent that there are irregularly shaped projectile fragments discontinuously distributed through the target thickness from the impact face to the exit face. Interestingly, there is an apparent scarcity of projectile fragments present toward the sample mid-thickness. When one looks closer at additional frontal section views such as

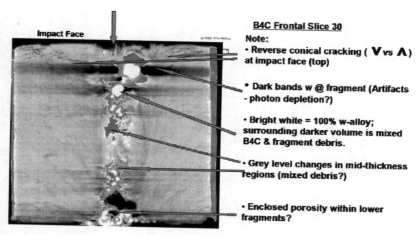

B4C Frontal Slice 30

Note:
• Reverse conical cracking (**V** vs **∧**) at impact face (top)

• Dark bands w @ fragment (Artifacts - photon depletion?)

• Bright white = 100% w-alloy; surrounding darker volume is mixed B4C & fragment debris.

• Grey level changes in mid-thickness regions (mixed debris?)

• Enclosed porosity within lower fragments?

Figure 2. Several observations are shown of features revealed in frontal slice F-30 of the B₄C ceramic target.

the non-segmented F-30 image shown in figure 2, an interesting observation is made – namely that additional fragments are now observed in the mid-section thickness that are less dense and have a considerable lower (darker) gray level than the bright white fragments which are essentially composed completely of tungsten alloy remnants. The medium lower gray level fragment volumes are still higher density than the B₄C ceramic and thus are indicative of an overall fragment density dilution due to the mixing of projectile and host ceramic debris. In fact, all of the fragments in this image appear to contain at least some of the medium density "corona" of mixed debris. The mixed density fragments appear discontinuous as do the other all projectile fragment segments. It is thus important to carefully consider the density segmentation range (mean + deviations in the selected gray level values) so that details like the above do not get overlooked.

Three additional observations illustrated in figure 2 are:
1. The presence of dark horizontal bands at target depths coincident with the highest density (white) projectile fragments. The fact that the apparent size of these dark bands increases as the size of the corresponding white fragment increases is strongly suggestive that the dark bands are in fact artifacts. They are most probably scanning artifacts caused by an insufficient photon fluence arriving at the linear detector array during the original XCT scanning due to the higher x-ray photon absorption by the tungsten alloy at the corresponding horizontal position.
2. The presence of "V" shaped cracking patterns with a convergent angle widest near the top impact surface. Typically, inverted "∧" or conical "hertzian" cracking is observed originating near the impact surface and becoming wider in diameter with increasing depth into the target sample. The later conical cracking is, however, observed originating near the bottom exit face and diverging inward.

3. The presence of some porosity (small irregularly shaped dark areas) is enclosed within the lower fragment mass near the exit surface. This is not unexpected since there are multiple smaller high density fragments surrounded by areas of ceramic and projectile debris agglomerated and intermixed in the near vicinity of the exit surface.

To measure various quantitative aspects of the individually selected fragments, a 3D morphological segmentation technique is invoked by placing a seed point within the fragment volume and using a region-growing algorithm to select all "fragment" material within the limits of a defined tolerance of voxel grey level values. The larger the tolerance value selected, the larger will be the detected fragment size with diluted density regions increasingly incorporated into the region of interest.

Eleven distinct embedded fragments containing an identifiable mass of all K.E. projectile material are identified in the B$_4$C ceramic target as shown in the virtually transparent frontal image on the right side of figure 3. Additional smaller such fragments are also present but were not investigated for this effort, nor were mixed fragments without a significant discernable K.E. projectile mass. Location and metrology data obtained for each of the selected eleven fragments are listed in Table II. The 3D location data is presented as the slice numbers of the orthogonal slices – sagittal (yz), frontal (xz), and axial (xy) slice planes respectively – located through the estimated center of each

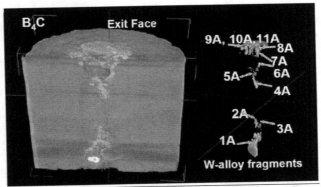

Figure 3. Segmentation and ID data are shown for multiple discrete projectile fragments embedded within the B$_4$C ceramic target

irregularly shaped fragment. The linear dimensions of the virtual spatial volume encompassing each fragment are given as the bordering dimensions. The actual surface area and the volume presented for each selected fragment were determined by software voxel measurement procedures. Variation in the magnitude of the values provided in Table II would be anticipated if somewhat different values of the segmentation tolerance

Table II- Location and Metrology Results for Projectile Fragments in B$_4$C Ceramic

Projectile Fragment ID	3D Location – slice# < Sag-Frontal-Axial>	Bordering Dimensions, mm	Surface Area mm^2	Volume mm^3
P-F #1A	< S212 F22 A84 >	14.82 x 4.51 x 19.12	416.30	264.32
P-F #2A	< S238 F29 A181 >	3.22 x 3.65 x 5.37	28.14	6.54
P-F #3A	< S252 F33 A172 >	54.18 x 7.20 x 37.14	7.42	1.38
P-F #4A	< S247 F22 A326 >	3.65 x 3.44 x 5.59	47.53	8.22
P-F #5A	< S239 F22 A342 >	1.50 x 0.86 x 1.29	4.17	0.60
P-F #6A	< S271 F22 A393 >	1.07 x 0.64 x 1.72	2.73	0.32
P-F #7A	< S268 F22 A440 >	4.94 x 5.59 x 9.02	119.41	26.47
P-F #8A	< S283 F22 A446 >	2.15 x 1.72 x 1.93	7.52	1.13
P-F #9A	< S191 F22 A444 >	2.36 x 1.93 x 1.50	9.13	1.27
P-F #10A	< S205 F33 A449 >	3.01 x 1.50 x 3.01	14.38	2.37
P-F #11A	< S279 F26 A450 >	2.15 x 1.72 x 1.93	7.35	1.08

parameter were employed. The ostensible size and morphology of any individual or agglomerated irregularly shaped fragment mass will also vary depending upon the visualization image in which that specific fragment mass is observed. For example, the largest individually segmented fragment (1A) is located near the impact surface, although casual visual inspection of figure 1 might suggest that the fragment mass closest to the exit face is larger. Closer inspection reveals that the fragment mass closest to the exit face is not a single homogeneous fragment but rather actually consists of several smaller fragments as shown in axial slice A-436 of figure 4. Some of these smaller individual fragments are labeled as 7A, 8A, 9A, 10A, and 11A in fig.3

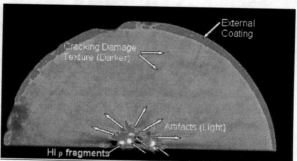

Figure 4. Axial scan A-436 revealing multiple smaller fragments within larger apparent fragment mass near the exit surface of the B$_4$C ceramic target.

The remaining salient observation to be presented made here for this B$_4$C target is the presence of inhomogeneous impact-induced porosity. As shown in figure 5, pores of varying size and morphology are located within and contiguous to the larger embedded fragments and also smaller pores are located away from the fragments in the surrounding epoxy coating. Similar observations of inhomogeneous impact-induced porosity are also presented subsequently for the impacted Al_2O_3 and TiB_2 targets as well.

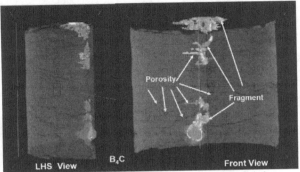

Figure 5. Transparent side and frontal views of the B_4C target reveal the presence of impact-induced porosity both within and contiguous to the larger fragment masses.

ALUMINUM OXIDE TARGET

The Al_2O_3 ceramic target has a large cavity with a depth to about 15 mm from the impact face. The target was fully penetrated to the rear surface as shown in figure 6. Although some limited fragment indications are observed in these two images, the fragment morphology and details are here indecipherable. However, further segmentation studies of the fragment volume revealed a series of discontinuous fragments extending through the entire remaining target thickness.

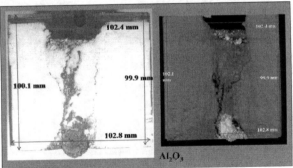

Figure 6. Full penetration was experienced by the Al_2O_3 target as shown by the macro-photograph (left) and the XCT virtual 3D solid object reconstruction (right).

A full longitudinal view of the fragments is evident in both the virtual 3D solid view and the fully transparent view of figure 7. This figure reveals a substantially larger local fragment mass concentration at the target exit face than in the remaining target thickness, while multiple small fragments are observed clustered near the base of the penetration cavity. In the interim thickness region, there is considerable irregularity in the morphology of the remaining fragments. The structural complexity of these fragment

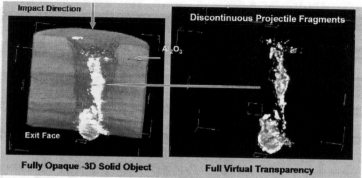

Figure 7. The irregular morphologies of the through thickness fragments are evident in both the 3D solid object view (left) and the full virtual transparent view (right).

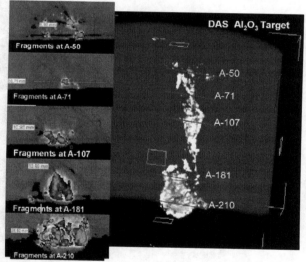

Figure 8. Close-up details of the residual fragments are shown for five representative virtual axial slices at the designated through thickness locations.

masses becomes more evident when viewed at higher magnification. Figure 8 reveals several magnified views of representative fragment details observed on the designated virtual XCT axial slice planes. The multiple higher density tungsten alloy fragments are observed surrounded with lower density material of intermixed debris from both the projectile material and the pulverized host ceramic material.

The considerable presence of inhomogeneous impact-induced porosity is also revealed in figure 9 for the Al$_2$O$_3$ target as seen previously above in the B$_4$C target. The size of each individual pore is indicated according to the assigned pseudo-color scale which ranges up to 250 mm^3. The radial distribution of these pores varies with the radial profile of the fragments to a maximum extent at the largest exit face fragment mass. Several of these observed pores are not contiguous with the fragment surfaces in the Al$_2$O$_3$ target, but rather are considerably distant. The porosity size distribution shown as a

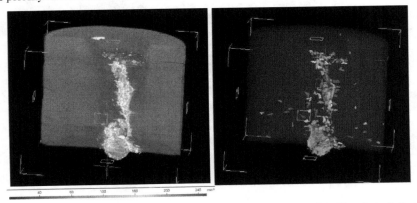

Figure 9. Impact-induced porosity near the through thickness fragments is evident in both the 3D solid object view (left) and the full virtual transparent view (right).

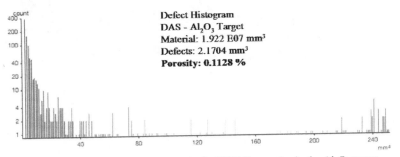

Defect Histogram
DAS - Al$_2$O$_3$ Target
Material: 1.922 E07 mm^3
Defects: 2.1704 mm^3
Porosity: 0.1128 %

Figure 10. A histogram of pore volumes found in the XCT diagnostics in the Al$_2$O$_3$ target

defect histogram in figure 10 reveals a wide distribution in the individual pore volume from a range <1 mm^3 to 250 mm^3. The total porosity volume was measured as 0.1128%.

TITANIUM DIBORIDE TARGET

The TiB2 ceramic target was only partially penetrated to about 60% of its thickness. A large cavity is observed on the impact surface. Two substantial and several small and discontinuous projectile fragments are observed in figure 11 as the virtual opacity of the target is reduced to fully transparency. One large primarily tungsten alloy fragment resides near the impact surface and the second lies in the approximate mid thickness of the target. Basic voxel metrology data for these two large fragments are presented in figure 12.

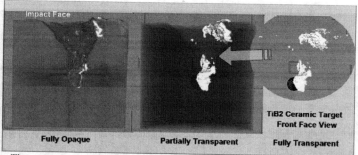

Figure 11. The presence and morphology of embedded projectile fragments are revealed through increasing virtual transparency in the TiB$_2$ ceramic target.

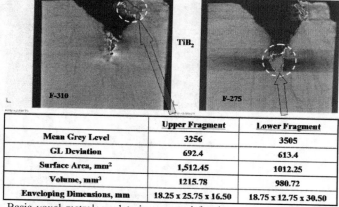

	Upper Fragment	Lower Fragment
Mean Grey Level	3256	3505
GL Deviation	692.4	613.4
Surface Area, mm^2	1,512.45	1012.25
Volume, mm^3	1215.78	980.72
Enveloping Dimensions, mm	18.25 x 25.75 x 16.50	18.75 x 12.75 x 30.50

Figure 12. Basic voxel metrology data is presented for the two large discontinuous projectile fragments embedded within the partially penetrated TiB$_2$ ceramic target.

It is interesting to note that the measured mean grey levels of the two large embedded fragments are substantially below the maximum grey level value of 4095 attributed to 100% tungsten-alloy. This is a firm indication of the intermixing of the projectile debris with ceramic debris to dilute the density (grey level) of the overall segmented projectile mass. An additional factor to consider is the 600+ level of the deviation from the measured mean grey level, which also reinforces the concept of variable projectile fragment material dilution. A further direct visualization of the density variations within and adjacent to the constituent segments of the large upper fragment mass is revealed by the grey level measurements shown in figure 13.

Figure 13. Grey level measurements observed on XCT axial slice A-167 indicate significant density variations in the upper fragment mass of the TiB_2 ceramic target.

In figure 14, several indications of inhomogeneous impact-induced porosity are observed in the TiB_2 ceramic target with their relative volumes distinguished by the applied pseudo-color. Most, but not all, of the observed impact-induced pores are contiguous to the fragment boundaries. The individual non-spherical pore shapes are observed to have relatively high aspect ratios as do the two large fragment masses.

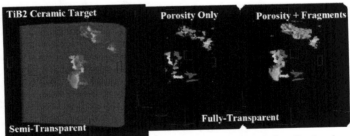

Figure 14. Impact-induced porosity of various sizes is shown mostly localized and contiguous to the two large embedded projectile fragments.

DISCUSSION
Several common aspects of the embedded fragments were revealed through XCT diagnostics despite the difference in the target materials. First, a wide range of ostensible fragment masses were observed in all three targets. Second, the overall fragment masses were found to be discontinuous and irregularly shaped. Third, an inner structural complexity of the larger fragment masses was revealed to consist of relatively small higher density tungsten alloy projectile remnants and a variable density mixture of finer debris from both the projectile and the host target ceramic. Fourth, the presence of inhomogeneous impact-induced porosity is present in each target in near proximity to the surface of the larger embedded fragment masses. Some

porosity is observed trapped within the fragment masses and also some pores are well removed from all fragments.

SUMMARY

The segmentation, metrology, 3D-visualization, and analysis of embedded ballistic fragments have been demonstrated using XCT diagnostics for three targets of quite different armor ceramic materials. The larger ostensible fragment masses are shown to contain an inner structural complexity not anticipated *a priori*. Essentially, small mostly high density tungsten alloy fragments (grey level of ~4095) are found surrounded by varying amounts of medium density material (grey level ~ 1800 to 4000) comprised of intermixed debris from both the projectile tungsten alloy and the pulverized host ceramic target.

Recommendations for further investigative efforts:

- Linkage of the mesoscale XCT fragment diagnostic results with future higher resolution interrogations by micro-tomography and electron microscopy.

- Exploring the relationships of the projectile fragments to the morphology of impact cracking in near proximity, and ultimately to their joint effect on penetration behavior.

- Development of similar fragmentation characterization capabilities for the host ceramic material.

ACKNOWLEDGEMENTS:

Grateful acknowledgements are extended to Dr. Donald A. Shockey for providing the author with access to the ceramic targets described herein and also, to Mr. William H. Green for his careful XCT scanning of these ceramic targets.

REFERENCES:

[1] J.M. Wells, "On the Role of Impact Damage in Armor Ceramic Performance", Proc. of 30th Intn'l Conf. on Advanced Ceramics & Composites-Advances in Ceramic Armor, v27, 7, (2006), *In Press.*

[2] D.A. Shockey, A.H. Marchand, S.R. Skaggs, G.E. Cort, M.W. Burkett and R. Parker, "Failure Phenomenology of Confined Ceramic Targets and Impacting Rods," *International Journal of Impact Engineering*, v 9 (3), 263-275, (1990)

[3] D.A. Shockey, D.R. Curran, R.W. Klopp, L. Seaman, C.H. Kanazawa, and J.T. McGinn, "Characterizing and Modeling Penetration of Ceramic Armor," ARO Report No. 30488-3-MS, (1995)

[4] D.A. Shockey, A.H. Marchand, S.R. Skaggs, G.E. Cort, M.W. Burkett and R. Parker, "Failure Phenomenology of Confined Ceramic Targets and Impacting Rods," Ceramic Armor Materials by Design, Ed. J.W. McCauley et. al., Ceramic Transactions, v134, ACERS, 385-402 (2002)

[5] Volume Graphics StudioMax v1.2.1, www.volumegraphics.com

[6] J. M. Wells, N. L. Rupert, and W. H. Green, "Progress in the 3-D Visualization of Interior Ballistic Damage in Armor Ceramics," Ceramic Armor Materials by Design, Ed. J.W. McCauley et al., Ceramic Transactions, v134, ACERS, 441-448 (2002).

[7] H.T. Miller, W.H. Green, N. L. Rupert, and J.M. Wells, "Quantitative Evaluation of Damage and Residual Penetrator Material in Impacted TiB₂ Targets Using X-Ray Computed Tomography," 21st International Symposium on Ballistics, Adelaide, Au, ADPA, v1, 153-159 (2004).

[8] J. M. Wells, "Progress on the NDE Characterization of Impact Damage in Armor Materials", Proceedings of 22nd International Symposium on Ballistics, Vancouver, ADPA, v1, 793-800 (2005).

[9] J.M. Wells and R.M. Brannon, "Advances in XCT Diagnostics of Ballistic Impact Damage," Dynamic Behavior of Materials, Edited by N. Thadhani, K. Vecchio, M. Meyers, G. Gray, and E. Cerreta, TMS, *2007, In Press*

[10] J.M. Wells, N.L. Rupert, W.J. Bruchey, and D.A. Shockey, "XCT Diagnostic Evaluation of Ballistic Impact Damage in Confined Ceramic Targets", 23rd International Symposium on Ballistics, Tarragona, Spain, ADPA v2, (2007) *In Press*

BALLISTICALLY-INDUCED DAMAGE IN CERAMIC TARGETS AS REVEALED BY X-RAY COMPUTED TOMOGRAPHY

H.T. Miller[*], W.H. Green, J.C. LaSalvia
U.S. Army Research Laboratory
Weapons and Materials Research Directorate
Aberdeen Proving Ground, MD 21005-5069

ABSTRACT
    X-ray computed tomography (XCT) has been shown to be an important non-destructive evaluation technique for revealing the spatial distribution of ballistically-induced damage in ceramics. However, the accuracy and level of damage that can be revealed by XCT imaging has not been completely determined. In this study, comparisons were made between the ceramic damage revealed using XCT images, and the damage observed optically within sectioned and polished cylinder cross-sections. Commercially-available armor-grade SiC cylinders (25.4 mm x 25.4 mm) were impacted with WC-6Co spheres (6.35 mm diameter) at velocities ranging from 200 m/s – 400 m/s. The recovered cylinders were then scanned using a 225 keV microfocus X-ray source. The X-ray images were evaluated using Volume Graphics© imaging software and various damage characteristics were observed. The cylinders were then sectioned and final polished at various distances from the center of impact. Damage comparisons between the XCT images and sectioned cylinders as a function of impact velocity will be presented.

INTRODUCTION
    Understanding the physical phenomena that govern the failure of ceramics under ballistically-induced impact is crucial in developing light-weight armor in the 21st century. Computational ceramic armor models have been the primary mode of investigating this problem since the early modeling work of Wilkins.[1] To garner better physical insight into the mechanisms surrounding damage in impacted ceramic targets, work by Hauver et al.[2] and LaSalvia et al.[3,4] have investigated the cross-sectional damage information in post-mortem recovered ceramic targets. These observations were important in identifying key damage characteristics, such as cone, radial, lateral, and median-vent cracks, as well as the comminuted region.
    Due to the three-dimensional nature of damage, Wells et al. have advocated using x-ray computed tomography (XCT) as a non-destructive technique in evaluating ballistic impact damage in candidate armor materials.[5,6] Non-destructive evaluation, via XCT, of ballistically-impacted ceramic targets has many benefits in visualizing damage features, and their role in ceramic armor performance. XCT allows for three-dimensional volumetric reconstruction of all damage features (within the limits of the spatial resolution of the particular XCT inspection procedure) in an impacted target. Furthermore, quantitative information can be obtained which can be utilized to improve modeling capabilities due to the physical confirmation provided by XCT. The objective of this investigation is to determine the extent at which XCT can confirm

[*] Work performed with support by an appointment to the Research Participation Program at the U.S. ARL administered by the Oak Ridge Institute for Science and Education through an interagency agreement between the U.S. Department of Energy and U.S. ARL.

the damage visualized in low velocity impacted ceramic targets, as well as to show some of the quantification ability that is available using this non-destructive technique.

EXPERIMENTAL PROCEDURES

Three armor-grade SiC-N cylinders, manufactured by Cercom, Inc., were used in this study. Material and mechanical property data for this material is located in Table I. All material and mechanical properties were determined using ASTM standards, and a discussion of these techniques can be found in previous work.[3]

Table I. Material Characteristics.

ρ (g/cm^3)	Grain Size (μm)	E (ν) (GPa)	HK4 (GPa)	K_{IC} (MPa*m$^{1/2}$)	σ_b (MPa)	Phases
3.22	3.3	452 (0.165)	18.6	5.1	500	6H, 15R, 3C

The cylinders were 25.4 mm in diameter and 25.4 mm in length. They were then mounted using Struers Isofast™ material which only exposed the front surface that would be impacted. The mounting material encased the ceramic cylinders in a 6.35 mm thick "Bakelite cup" around the sides and back of the specimen. The encasing of the ceramic in this manner was done for three reasons. First, mounting the ceramic would provide some support to hold the specimen together when it was impacted at the higher velocities. Second, this would be a less attenuating material to be XCT scanned through, and generally result in better quality images. Finally, sectioning the samples would not be as difficult as opposed to using a Ti-6Al-4V cup, which was used in previous work.[3,4]

The mounted targets were impacted with a WC-6Co sphere (nominal diameter of 6.35 mm) at velocities of 207 m/s, 318 m/s, and 385 m/s. The density and hardness of the spheres, as measured by Wereszczak, were 14.93 g/cm^3 and 16.3 GPa (HV0.5), respectively.[7] The WC-6Co spheres were launched from a foam sabot, having a diameter of 25.4 mm. The sabot was completely detached from the sphere by using a stripper plate to remove the sabot prior to impact with the ceramic. A more thorough description of the ballistic test setup and procedures is described elsewhere.[8]

The recovered cylinders were left in their mounts and inspected using XCT. They were scanned in rotate-only (RO) mode using the 420 keV/225 keV XCT inspection system at the U.S. Army Research Laboratory. In RO mode the object being scanned rests on a rotating turntable, which goes through one full rotation (360°) with the object remaining completely within the horizontal (x-ray fan beam) field of view (FOV). The x-ray source and detector used were the 225 keV microfocus tube and the image intensifier/1024^2 CCD camera combination, respectively. The energy and current of the microfocus tube were 190 keV and .040 mA, respectively, with a focal spot size of approximately 5 μm. The source-to-object-distance (SOD) and source-to-image-distance (SID) were 155 mm and 645 mm, respectively, resulting in a magnification factor of 4.16. The FOV, slice (scan) thickness, and vertical slice increment were 45 mm, .245 mm, and .200 mm, respectively. To completely inspect each sample required three hours of scanning time, approximately 130 slices, with each scan (slice) collecting 1600 views (projections), which were reconstructed to a 1024 x 1024 image matrix.

Following the XCT scanning of the recovered cylinders, the samples were each sectioned in three locations perpendicular to their impact faces. An initial cut was made approximately 2 mm away from the center of impact (this would later be polished down to the center of impact). Two subsequent cuts were made on either side of the initial cut at different locations in each sample. This resulted in four axial views in each of the three samples. The axial sections were metallographically prepared using 15 μm, 6 μm, and 1 μm diamond slurries. Careful attention was made in measuring how much material was removed during each step. This would be crucial in accessing the precise location that is being observed in the comparative three-dimensional reconstructed XCT axial sections.

RESULTS AND DISCUSSION
The three recovered sphere impacted SiC-N cylinders are shown in Figure 1. Concentric ring cracking and radial cracking were observed in all of the samples. Significant damage was observed in the 318 m/s and 385 m/s impacted specimens. In these samples, radial cracking extended through the "Bakelite cup", limiting the ability of this technique to go beyond 400 m/s. In the 385 m/s sample, large sections of material were removed upon impact with the WC-6Co projectile.

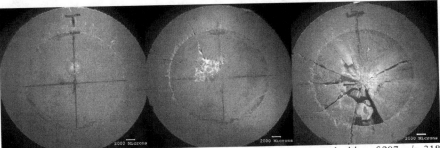

Figure 1. From left to right, the recovered impacted ceramic targets at velocities of 207 m/s, 318 m/s, and 385 m/s, respectively.

XCT Single Scans and Three-Dimensional Visualization
The XCT scans of the impacted ceramic targets revealed many damage features that could not otherwise be seen. Figure 2 shows cross-sectional XCT scans of the 385 m/s sample, parallel to its face at various depths from the top (impact) face. Outlined by the dotted line in these scans is the boundary between the ceramic sample and the Bakelite mounting material. Radial cracking can be seen extending beyond the ceramic/Bakelite interface in all four scans. The depth of the large sections of material that were removed upon impact is shown to reach at least 1.6 mm, but does not go beyond 3.6 mm. The growth of the primary cone crack can be seen intersecting radial cracks, and nearly reaching the diameter of the sample at a depth of 10.4 mm. The small dark spot that appears at the center of a few of the images is an artifact due to an uncorrectable irregularity in the image intensifier.
Cross-sectional XCT scans of the 318 m/s sample, parallel to its impact face, are shown in Figure 3. The level of discernable damage in this sample decreases dramatically when

compared to the level of damage in the 385 m/s impacted sample. Radial cracking extending from the cone crack and through the Bakelite cup can be seen. It is important to note the depth at

Figure 2. From left to right, cross-sectional XCT scans of the 385 m/s sample (parallel to its impact face) at 1.6 mm, 3.6 mm, 8.4 mm, and 10.4 mm from the impacted face, respectively.

Figure 3. From left to right, cross-sectional XCT scans of the 318 m/s sample (parallel to its impact face) at 0.8 mm, 1.8 mm, 4.8 mm, and 5.8 mm from the impacted face, respectively.

which the cone crack reaches the ceramic/Bakelite interface. In this case, at a depth of 5.8 mm the cone crack has expanded to the diameter of the cylinder. This is considerably different than in the 385 m/s impacted sample which still has a fully established cone crack at 10.4 mm. Thus, the cone angle of the 385 m/s sample is less than that of the 318 m/s sample. The difference is attributed to the decreased impact velocity, and the shallower cone crack angle that forms as a result.[3,4] Again, the small dark dot that appears at the center of the images is an artifact, due to an uncorrectable irregularity in the image intensifier.

The sample impacted at 207 m/s did not show any easily discernable damage in the XCT scans, and therefore scans are not shown. Image contrast techniques were used in an attempt to improve visualization of cracking in this sample. However, although this approach extracts faint indications of similar types of damage as in the other samples, they do not particularly stand out from the surrounding background gray level variations. Secondly, the width of the significant majority of the cracking in this sample appears to be below the limit of the spatial resolution of this particular XCT parameter approach.

Volume Graphics© three-dimensional imaging software was used to volumetrically "stack" and digitally section the XCT scans to compare them to the physical metallographic cross-sections. Figure 4 shows the XCT surfaces and the corresponding mechanically polished surfaces at the center of impact, 1.7 mm from the center of impact, and 6.0 mm from the center of impact in the 385 m/s sample. The XCT surfaces show reasonably good agreement with the metallographically prepared samples. The cone crack can be observed as well as the bifurcation of this crack in both the 1.7 mm and 6.0 mm distant surfaces. The larger radial cracks are also seen extending horizontally through the cylinder. The finer cracks near the impact site can not be individually seen, in part due to the resolution limit of the approach, but also because of the artifact problem that occurred during these scans. It is also generally apparent that the cracking is wider and more severe on the left hand side of the XCT surfaces. Secondly, it qualitatively appears that the level of cracking from the impact surface down towards the bottom, or "vertical" cracking, is more severe in the left hand side of the metallographic surfaces. It is reasonable to postulate that the more severe level of this cracking may have allowed interconnected conical and/or radial cracks to "open up" more, thus becoming wider and/or larger. Lastly, the slightly upward curved horizontal cracking towards the bottom of the metallographic surfaces does not appear to show up at all in the XCT surfaces. This may be because they are simply too tight to be resolved in the XCT scans or they may be a result of the physical sectioning. It is not possible to definitively know which of these possibilities is the case.

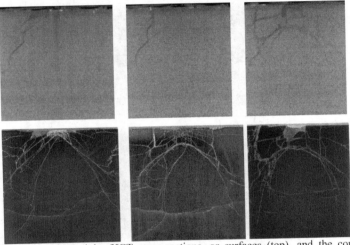

Figure 4. From left to right, XCT cross-sections, or surfaces (top), and the corresponding mechanically polished surfaces (bottom) at the center of impact, 1.7 mm away from the center of impact, and 6.0 mm away from the center of impact, respectively, in the 385 m/s sample.

Figure 5 shows the XCT surfaces and the corresponding mechanically polished surfaces at the center of impact, 1.7 mm from the center of impact, and 5.1 mm from the center of impact in the 318 m/s sample. The XCT surfaces show fair agreement with the metallographically prepared samples, with the more severe cracking appearing on the right hand side of the XCT surfaces. Again, this may be due to the presence of cracks coming down from the impact surfaces allowing interconnected cracks to become wider and/or larger. Also, the "cupped" horizontal crack on the left towards the bottom of the polished surface at the center of impact does not appear to show up in the corresponding XCT surface.

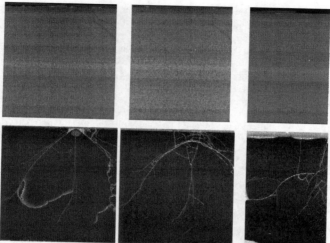

Figure 5. From left to right, XCT cross-sections, or surfaces (top), and the corresponding mechanically polished surfaces (bottom) at the center of impact, 1.7 mm away from the center of impact, and 5.1 mm away from the center of impact, respectively, in the 318 m/s sample.

Determination of the Size of a Cone Crack

The ability to acquire quantitative damage information is one of the most attractive features in utilizing XCT evaluation to provide data for computational modeling comparison or validation. Figure 6 shows the steps of an image analysis technique which allows for the determination of cone crack diameters through the thickness of a sample. In this case, the images are of the 385 m/s sample. Figure 6a is a highly contrasted XCT slice (image) in the 385 m/s sample, in which standard gray level windowing has produced the desired effect. In Figure

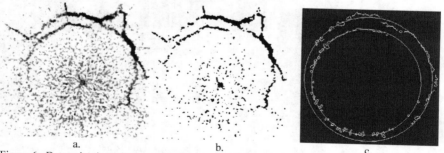

a.   b.   c.

Figure 6. Determination of cone crack diameter in an XCT slice (385 m/s sample). a) highly contrasted XCT slice b) segmented binary XCT slice c) circle fit to determine diameters

6b, this has been taken to the binary limit, which produces a completely segmented black and white image. Figure 6c is generated by a three step process. First, XCT scanner software creates what is called a point cloud by locating points in three-dimensional space along all of the boundaries between black and white in the binary segmented image (Figure 6b). Second, the data points of interest are parsed out of the initial set of points by another appropriate software application. Third, a circle is mathematically fit to the parsed set of data points that represent the physical path of the cone crack. In fact, this can be done in any slice in which the cone crack is defined enough to determine its path. The images shown in Figure 6 from slice number 220 (see Table II) show two cone cracks and their mathematical fits, thus exhibiting the nature of the bifurcation of the cone crack seen in the 385 m/s sample.

Table II shows the radii, centers, and depths of the cone crack in the 385 m/s sample for several slices starting near the impact face. It can be seen that after removing slice number 250 as a possible outlier, the center of the cone crack remains in approximately the same (x,y) planar location with increasing depth. Secondly, slices 220 (see Figure 6) and 230 show data for both

Table II. Radii, Centers, and Depths for Primary Cone Crack in 385 m/s sample

Slice Number	Radius (mm)	Center (x,y) (mm)	Depth (mm)
196	2.00	(-0.65, 1.09)	1.6
197	2.12	(-0.81, 1.10)	1.8
198	2.48	(-0.59, 1.06)	2.0
199	2.58	(-0.66, 1.08)	2.2
200	2.78	(-0.73, 1.37)	2.4
201	2.92	(-0.74, 1.23)	2.6
202	3.13	(-0.70, 1.24)	2.8
203	3.27	(-0.62, 1.19)	3.0
206	3.77	(-0.78, 1.21)	3.6
211	4.57	(-0.69, 1.17)	4.4
220	5.77 (1) 6.76 (2)	(-0.92, 1.24) (1) (-0.71, 1.55) (2)	6.4
230	6.93 (1) 8.91 (2)	(-1.14, 1.22) (1) (-0.79, 1.02) (2)	8.4
240	7.81	(-1.31, 1.09)	10.4
250	9.01	(-0.50, 0.94)	12.4

parts of the bifurcated cone crack. Figure 7 shows the plot of radius vs. depth for the primary cone crack in the 385 m/s sample, along with two additional data points for the bifurcated part. The angle of the primary cone crack can also be determined from the radius and depth data in Table II and is 66°. It is noted that this plot shows similarities with the XCT and physical metallographic surfaces shown in Figure 4.

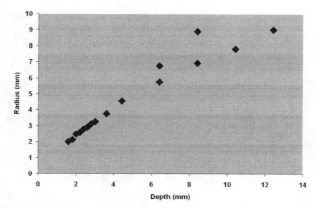

Figure 7. Plot of radius vs. depth of the primary cone crack in the 385 m/s sample.

SUMMARY AND CONCLUSIONS

A set of three 25.4 mm diameter by 25.4 mm thick ceramic SiC-N cylinders mounted in Bakelite cups were each impacted with a WC-6Co sphere at velocities between 200 m/s – 400 m/s. Each of the impacted cylinders was visually inspected followed by a set of x-ray computed tomography scans in order to characterize the damage. The scanned cylinders were then physically sectioned, perpendicular to their impact faces, in three locations. These sections were then metallographically prepared to reveal damage information at those surfaces.

In evaluating the damage, individual cross-sectional (parallel to faces) XCT slices, three-dimensional reconstructed volumetric XCT images, and physically sectioned specimens were used. The individual XCT slices (images) showed significant cracking damage in the 385 m/s and 318 m/s samples. The evolution of the cone crack in both of these specimens is also easily observable through the thickness (i.e., depth) of the cylinder. Individual XCT slices were volumetrically "stacked" using Volume Graphics© software, and surface views through the specimen from top to bottom were observed. These surface views were compared to the metallographically prepared specimens at approximately the same locations to discern the ability of XCT to capture key damage features. In comparing the XCT and real surfaces, XCT showed reasonably good agreement with the polished samples in exhibiting similar damage features in the material. Relatively fine micro-damaged regions and possibly some tight cracking damage could not be identified due to the combination of the resolution limit of the particular XCT parameter approach used and the presence of artifacts in some of the images. However, two important points are noted here. First, it should be possible to improve on image quality in future work. Second, it is not always necessary to capture every detail of all damage features to obtain and evaluate useful and important qualitative and/or quantitative information about a sample, as shown in this paper.

The ability to quantify damage features using image segmentation techniques on XCT scans was shown. The radii, centers, and depths of the cone crack in the 385 m/s sample were calculated using these techniques. The center of the cone crack remained in approximately the same (x,y) location and the plot of radius vs. depth of the cone crack looked graphically similar to features that were present in the XCT and physical metallographic surfaces of the sample. This is a powerful approach to determining useful and potentially important quantitative information about ballistic damage. Quantifiable damage information is desirable in damage modeling to confirm the features as well as to garner new insight into damage that has not been captured in the models. The full capabilities of the XCT diagnostic approach have not yet been reached and the beneficial utilization of this new volumetric damage knowledge has yet to be extensively applied and exploited. Secondly, this new volumetric damage knowledge has yet to be utilized in ballistic damage models for comparison to the models and subsequent analysis. It is highly desirable to develop the XCT diagnostic approach along these two paths to create an efficient, useful, and powerful investigative methodology for characterizing, analyzing, and comparative modeling of volumetric shock/impact damage.

REFERENCES
[1]C.E. Anderson, "A Review of Computational Ceramic Armor Modeling," *Cer. Eng. Sci. Proc.*, **27**[7] (2006).
[2]G.E. Hauver, E.J. Rapacki, P.H. Netherwood, and R.F. Benck, "Interface Defeat of Long-Rod Projectiles by Ceramic Armor", ARL Technical Report, ARL-TR-3590, September 2005, 85 pp.
[3]J.C. LaSalvia, M.J. Normandia, H.T. Miller, and D.E. MacKenzie, "Sphere Impact Induced Damage in Ceramics: I. Armor-Grade SiC and TiB$_2$," *Cer. Eng. Sci. Proc.*, **26** [7] 171-179, (2005).
[4]J.C. LaSalvia, M.J. Normandia, H.T. Miller, and D.E. MacKenzie, "Sphere Impact Induced Damage in Ceramics: II. Armor-Grade B$_4$C and WC," *Cer. Eng. Sci. Proc.*, **26** [7] 180-188, (2005).
[5]J.M.Wells, N.L. Rupert, and W.H. Green, "Visualization of Ballistic Damage in Encapsulated Ceramic Armor Targets," 20th International Symposium on Ballistics, v2 Terminal Ballistics, eds. J. Carleone and D. Orphal, DEStech Publications, Lancaster, PA, 2002, 729-738.
[6]J.M. Wells, W.H. Green, N.L. Rupert, J.R. Wheeler, S.J. Cimpoeru, and A.V.Zibarov, "Ballistic Damage Visualization & Quantification in Monolithic Ti-6Al-4V with X-ray Computed Tomography," 21st International Symposium on Ballistics, DSTO, Adelaide, Australia, ADPA, v1, pp. 125-131, 2004.
[7]A.A. Wereszczak, "Elastic Property Determination of WC Spheres and Estimation of Compressive Loads and Impact Velocities That Initiate Their Yielding and Cracking," *Cer. Eng. Sci. Proc.*, to be published.
[8]M.J. Normandia, D.E. MacKenzie, B.A. Rickter, and S.R. Martin, "A Comparison of Ceramic Materials Dynamically Impacted by Tungsten Carbide Spheres," *Cer. Eng. Sci. Proc.*, **26**[7], 2005.

ON CONTINUING THE EVOLUTION OF XCT ENGINEERING CAPABILITIES FOR IMPACT DAMAGE DIAGNOSTICS

J.M. Wells, Sc.D.
Principal Consultant, JMW Associates, 102 Pine Hill Blvd., Mashpee, MA 02649
(774)-836-0904   jmwconsult1@comcast.net

ABSTRACT

The continuing evolution of ballistic damage diagnostic techniques utilizing X-ray computed tomography, XCT, is providing new knowledge about the mostly unexplored and complex physical details of ballistic impact damage. While developmental results to date are impressive, further improvements in the non-invasive XCT damage diagnostic capabilities are required for a more complete volumetric characterization and engineering analysis of actual ballistic impact damage features. Despite the availability of substantial XCT scanning capabilities in the industrial and government sectors, the application of such capabilities toward the diagnostic evaluation and analysis of ballistic impact damage has been minimal to date. The ultimate functionality of improved impact damage characterization knowledge will be maximized with enhanced interactive collaboration between experimentalists, materials and NDE specialists, structural mechanics experts, and computational modelers. Utilization of such impact damage knowledge can be applied to the development, validation, and verification of damage-based computational modeling and simulations and ultimately to facilitate decision making related to the development of improved damage tolerant materials. Several XCT diagnostic developments envisioned by the author are described in the near-term (2008), mid-term (~ 2012), and far-term (~ 2020). The purpose of this presentation is to identify further potential impact damage diagnostic developments and their technical challenges for improved capabilities and effective application with damage-based ballistic computational modeling.

INTRODUCTION

The initial developments and substantial results of ballistic impact damage diagnostics with x-ray computed tomography, XCT, have been described by the author and his collaborators [1-11]. While much still remains to be done to further improve and refine the functionality of this non-invasive diagnostic technology, several unique and substantial capabilities have already been demonstrated including:

- Completely non-invasive volumetric digitization of both metallic and ceramic impacted terminal ballistic targets.
- 3D rendering of high resolution virtual solid object target reconstructions
- Virtual sectioning of such targets on orthogonal and arbitrarily oriented planes revealing multiple impact damage features.
- Segmentation and metrology of residual projectile fragment sizes, morphologies, intermixed sub-structure, metrology, and spatial distributions.
- Quantification and 3D visualization of inhomogeneous impact-induce porosity (void) distributions.
- Examination of multiple impact cracking modes in-situ and the detection of spiral and hourglass shaped cracking modes previously unreported in the ballistic literature.

- Impact surface cracking modes and topography changes, and irregular subsurface distortions.
- Axisymmetric quantification and 3D mapping of mesoscale impact cracking and fragmentation damage.

Remarkably, and quite unlike the strong development and ubiquitous application growth experienced with medical tomographic and digital imaging capabilities over the past decade, the XCT ballistic damage diagnostics technology has been only sparsely supported, practiced, or adopted within the terminal ballistic experimental, analytical, and/or computational modeling communities. Considerable effort continues to be extended in the development of other advanced visualization techniques such as real-time high speed photography [12-14] and more traditional NDE modalities such as ultra sound [15], for the purpose of ballistic impact damage diagnostics. Unfortunately, no integrated activity is apparent to date to combine these various non-invasive techniques into a more coordinated, quantitative, and holistic diagnostic approach for ballistic impact damage characterization, visualization, and analysis. Furthermore, the desired incorporation of such damage diagnostic capabilities have yet to be effectively utilized in the development, validation, and verification of damage-based analytical or computational models for ballistic performance. In this paper, the author suggests further improvements considered technically feasible and realistically anticipated in the continuing evolution of XCT impact damage diagnostic capabilities. Also briefly discussed are impact damage diagnostic applications and some potential ways in which they might be synergistically utilized and coordinated with other damage visualization techniques.

BACKGROUND

The purpose of an ideal ballistic impact damage diagnostic evaluation is to locate, identify, characterize measure, analyze, and spatially visualize all of the various physical impact damage features created in a defined target by the specified impact event. Non-invasive virtual diagnostic methods are most desirable in that they avoid the introduction of extrinsic physical damage features during destructive sectioning and also avoid the irreversible physical destruction of the ballistic target. By using virtual XCT diagnostic and analysis techniques, one can repeatedly reset the damage interrogation parameters with impunity.

One drawback of the mesoscale industrial XCT approach is the limitation on minimum feature resolution for substantial size ballistic targets. Depending on the encircling diameter and the density of the ballistic target, typical resolution levels of ~ 0.1 to 0.25 mm are achievable. Higher levels of resolution to ~10 micrometers are obtainable with micro-focus tomographic interrogations but these require somewhat smaller target size limitations. Still higher sub-micron resolutions are possible with synchrotron facilities but have such small sample size limitations as to make examination of intact laboratory ballistic targets of interest impractical. The destructive sectioning of actual larger ballistic targets subsequent to mesoscale XCT scanning and analysis remains an option for metallographic examination or interrogation of reduced portions of the original target with micro-tomography approaches. Promising real time XCT diagnostic capabilities are under development at Fraunhofer-EMI [16, 17] but have current resolution limitations of ~1 mm, substantially greater than that of static industrial micro- or even meso-scale XCT.

Unfortunately, there is presently no broadly accepted definition of ballistic impact damage. This is particularly troubling since the diverse ballistic community participants need to agree on what constitutes impact damage in order to proceed with appropriate damage measurements as well as approaches to model it. An excellent review of the current status of computational ceramic armor modeling was presented at this conference last year by Anderson [18]. Frequently, computational modelers, starting back with Wilkins in 1967 [19], invoke an internal state damage tracking parameter within a computational cell that varies from a value of 0 (undamaged) to a value of 1 (completely damaged or locally fractured). Such a damage parameter is essentially a mathematical cumulative plastic strain indicator over an adjustable volume element which is not indicative or descriptive of the multiplicity of discrete and complex 3-D physical damage features occurring in real ballistic impact conditions. Experimentalists generally seem to focus on impact-induced 2-D cracking damage manifestations - especially of radial, ring, conical, and laminar morphologies – all of which are readily observable at variable magnification with destructive and irreversible metallographic sectioning.

The present author prefers a much broader definition of ballistic impact damage [6] as: diverse irreversible impact-induced physical modifications of the material shape, structure, continuity, and/or structural integrity resulting from one or more specific ballistic impact event(s). This considerably broader definition of impact damage thus includes the following XCT observable and volumetric damage features:

- Discontinuous embedded residual projectile fragments,
- Host target material fragmentation,
- Multiple micro- and mesoscale cracking morphologies,
- Impact-induced porosity,
- Surface cratering and morphological changes resulting from projectile/ceramic ruble mixing, and
- Inhomogeneous target distortion or deformation.

Figure 1. presents representative examples of some impact damage features captured with the XCT diagnostic modality versus the 2D ceramographic sectioning and microscopy. The detection, spatial location, size, morphology, and metrology of the various distinct damage features are important to more fully characterize the "total physical damage state". Damage features occur on a continuum of length (1D), area (2D) and volume (3D) scales. To obtain a true "damage state characterization", one would need to accomplish this characterization across multiple "length" scales from the sub-micron to the mega-scale. Obviously, such a comprehensive volumetric interrogation is extremely difficult for a strictly non-invasive approach on substantial sized targets and is not normally conducted. Rather, diagnostic damage studies are typically conducted either on the micro-scale or on the meso-scale, but infrequently on both. Consequently, the linkage of complex damage features from the very small (< 10 microns) to the substantially larger (>10 mm) is seldom established, even though both simultaneously exist even within an impacted, but non-penetrated armor ceramic.

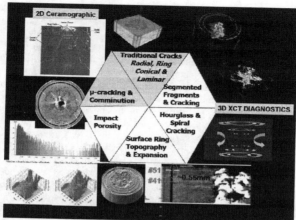

Figure 1. Comparative overview of various 3D impact damage features detected and visualized with XCT versus 2D ceramographic observations.

EXTENSION OF XCT DAMAGE DIAGNOSTIC CAPABILITIES

While developmental results to date are impressive, further improvements in the non-invasive XCT damage diagnostic capabilities are required for a more functional and complete volumetric characterization and engineering analysis of actual ballistic impact damage features. The improved characterization of impact-induced damage features is a necessary but not a sufficient condition for understanding the relative significance of such features. It is thus desirable that such further improvements be linked to the specific requirements of evolving computational ballistics models that are attempting to incorporate impact-induced damage in addition to penetration considerations [20, 21]. The comparison of model predictions of internal damage against the high resolution XCT visualizations and metrology of actual internal damage features is preferable to merely assessing a model's ability to predict penetration depth. This more stringent and physically significant validation of internal damage morphology is especially important if one hopes to perform subsequent "second strike" analyses. It is quite likely that most of the observable impact-induced damage features contribute to the degradation of the structural integrity of the target material and thus enable subsequent penetration. Their relative importance in the process of enabling penetration remains to be deciphered, as does the deduction of feasible ways in which to impede or mitigate their respective penetration enabling contributions. It is possible that some such features, although created by the impact event, are neither associated with hindering nor enhancing the initial strike penetration process. Damage features created by an initial impact may have a significantly different influence on the target behavior experiencing secondary impacts. Thus it is important to distinguish, clarify, and ultimately prioritize which such observed impact-induced damage features have the most significant deleterious effects on the overall ballistic performance of the target.

The development of further XCT diagnostic capabilities could be effectively conducted concurrently and collaboratively with the development of damage-based computational modeling. Not only would the diagnosed damage features become increasingly available in a more useful format, but also the model-predicted damage could be directly validated and verified by the non-invasive XCT interrogation of experimentally impacted targets. A schematic diagram highlighting the broad linkages of such collaboration is shown in figure 2.

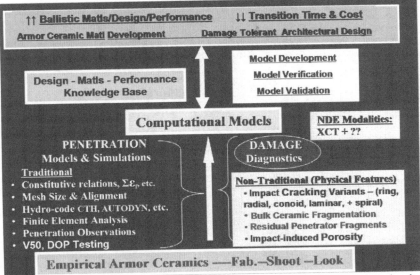

Figure 2. Schematic showing overall linkages of damage and penetration to computational ballistic performance modeling.

FUTURE REQUIREMENTS

Among the many aspects of the diagnostics and analysis of ballistic impact damage that remain to be improved upon are included:

- Improved feature segmentation quality especially of complex cracking morphologies.
- Automated techniques for the characterizing of multiple imbedded projectile fragments.
- Characterization and metrology of host ceramic fragmentation details.
- Increased utilization of higher resolution microfocus, XMCT, diagnostic capabilities.
- Linkage of micro- and meso-characterization of impact damage features.
- Improved speed and ease of damage feature segmentation and metrology.
- Automation techniques for quantification and 3D mapping of asymmetric damage.
- Techniques for the deconvolution of complex intermixed damage modes.
- Expansion of experimental damage diagnostic knowledgebase to include:
  - Transparent Materials (Glass, ALON, Spinel, Polyurethane, other?)
  - Additional interface defeat target samples.
  - Variable velocity impact tests with a broader range of target materials.
  - Linkage of damage features from indentation tests to ballistic impact tests.

- Pre-existing damage effects on secondary impacts (multiple hit behavior).
- Multi-slice and volumetric XCT scanning techniques.
- Stereo & animated (Fly-Through) visualizations of in-situ damage states.
- Multi-Modality with other diagnostic techniques (Microscopy, Ultrasound, EOI, etc.)
- Linkage of damage features to damage mechanisms.
- Linkage of damage features to penetration mechanisms.
- Assessment of relative significance of damage features.
- Guidance for improved material damage tolerance capabilities.
- Material modification guidance from damage characterization.
- Knowledgebase management, KBM, of impact damage for interactive access.
- Others……..

A preliminary overview of the suggested chronology for the concurrent XCT damage diagnostic evolution and the application of such capabilities within impact damage-based computational model development is presented in figure 3. Such a projection is obviously subject to revision depending upon many factors relating to the motivation, level of collaboration, funding, and participation of the necessary expertise from the experimental ballistic testing, materials processing and development, NDE characterization, and computation ballistics communities.

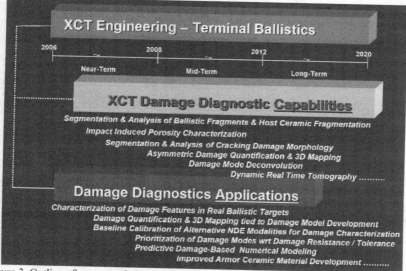

Figure 3. Outline of a proposed chronology for the development of selected XCT diagnostic capabilities and related damage diagnostic applications.

ON THE NATURE OF VISUALIZATION AND PERCEPTION

Understanding the form and nature of various physical characteristics of impact damage observed in an image is often not a simple exercise. Real or virtual images may contain non-real features (artifacts) that obfuscate the understanding of real object details being observed. Understanding basically involves the combined synergy of visualization and perception. By visualization here we mean being able to create a cognitive mental, as well as physical, image representative of the shape or form and sub-structural details of the actual object in 3D space. Perception here signifies the ability for acute and intuitive cognition of the relationships between complex and intermixed observed image features and a capacity for their discernment, discrimination and comprehension with respect to the ostensible image visualization. Advances in modern image processing technology have greatly improved and expanded the capability to analyze as well as capture volumetric image content. Such capabilities are available in the Volume Graphics StudioMax v1.2.1 advanced voxel analysis and visualization software [22] utilized for the XCT impact damage diagnostics discussed here. When assessing the validity of various image features, it also helps to utilize one's intuition and imagination in the context of the known facts and one's experience with similar circumstances.

IMPORTATION OF DAMAGE KNOWLEDGE INTO COMBINED MODELS

The XCT diagnostic modality exhibits the unique ability to perform in-situ impact damage characterization, analysis, and 3D visualizations in ways that no other NDE diagnostic modality has demonstrated. Consequently, it holds the exciting potential for identifying, characterizing, and visualizing all of the most significant damage morphologies occurring under diverse ballistic impact conditions. The impact damage puzzle is slowly being deciphered and will eventually be imported into damage-based modeling as shown schematically in figure 4

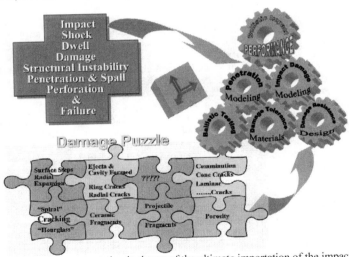

Figure 4. A schematic representation is shown of the ultimate importation of the impact damage puzzle into a combined penetration and damage-based modeling approach.

DISCUSSION

The challenges involved in incorporating physical impact damage knowledge into evolving damage-based computational modeling efforts are difficult and not to be understated. An initial exploration of the difficulties involved, and also the benefits to be derived from their resolution, was addressed earlier [4, 6]. Certainly, the broad ballistics community needs both a common physically based definition of impact-damage and robust damage-based analytical and/or computational modeling approaches to predict and understand the complex physical impact damage conditions actually experienced. The XCT diagnostic technology has been demonstrated to be the most effective non-invasive tool for in-situ damage characterization and visualization to date. Hopefully, improvements in other NDE modalities will be achieved to permit their synergistic contributions to the detection and characterization of ballistic impact damage. When adequate damage-based computational models become available, their validation and verification will be greatly assisted with XCT damage diagnostic results.

Additional evolutionary refinements and improvements in the XCT diagnostic capabilities for ballistic impact damage have been identified and are realistically anticipated to become available. The time frame for their availability and functionality will realistically depend upon the interactive support and technical collaboration of the ballistic experimental, NDE, and computational modeling communities.

"A pessimist sees the difficulty in every opportunity; an optimist sees the opportunity in every difficulty." - Winston Churchill

SUMMARY

While developmental results to date are impressive, further improvements in the non-invasive XCT damage diagnostic capabilities are both required and feasible for a more functional and complete volumetric characterization and engineering analysis of actual ballistic impact damage. The development of further XCT diagnostic capabilities could be most effectively conducted concurrently and collaboratively with the evolving development of damage-based computational modeling. Several specific XCT diagnostic technique improvements have been identified and an initial attempt at a development chronology has been proposed. Eventually, when effectively combined with advanced damage-based computational modeling, a number of innovative damage tolerant material structures and target architectures may be identified, designed, evaluated, and validated for a broad range of relevant armor applications. Exploring the implementation of XCT in conjunction with physical damage-based computational modeling and simulation for such applications requires still further enhancements in the improved resolution, segmentation, metrology, and 3D visualization of impact damage features, as well as an improved understanding of their significance and potential mechanisms to mitigate, redistribute, and/or prevent the initiation and growth of such damage features.

ACKNOWLEDGEMENTS

Grateful acknowledgements are extended to Drs. David Stepp, Doug Templeton, Leo Christodoulou, Bill Bruchey, Steve Cimpoeru, Don Shockey, Rebecca Brannon, Mr. Nevin Rupert and Bill Green for their advice and encouragement in the furtherance of the pursuit of the role of impact damage on the performance of armor ceramics.

REFERENCES

[1] W.H. Green, and J.M. Wells, "Characterization of Impact Damage in Metallic / Non-Metallic Composites Using X-ray Computed Tomography", AIP Conf. Proc. 497, 622-629 (1999).

[2] J. M. Wells, N. L. Rupert, and W. H. Green, "Progress in the 3-D Visualization of Interior Ballistic Damage in Armor Ceramics," **Ceramic Armor Materials by Design**, Ed. J.W. McCauley et al., Ceramic Transactions, v134, ACERS, 441-448 (2002).

[3] H.T. Miller, W.H. Green, N. L. Rupert, and J.M. Wells, "Quantitative Evaluation of Damage and Residual Penetrator Material in Impacted TiB$_2$ Targets Using X-Ray Computed Tomography," 21st International Symposium on Ballistics, Adelaide, Au, ADPA, v1, 153-159 (2004).

[4] J.M. Wells, "On Incorporating XCT into Predictive Ballistic Impact Damage Modeling," 22nd International Symposium on Ballistics, Vancouver, Ca, , ADPA, v1, (2005).

[5] J. M. Wells, "Progress on the NDE Characterization of Impact Damage in Armor Materials," Proceedings of 22nd International Symposium on Ballistics, Vancouver, ADPA, v1, 793-800 (2005).

[6] J.M. Wells, "On the Role of Impact Damage in Armor Ceramic Performance", Proc. of 30th Intn'l Conf. on Advanced Ceramics & Composites-Advances in Ceramic Armor, v27, 7, (2006), *In Press*.

[7] W.H. Green, N.L. Rupert, and J.M. Wells,, "Inroads In The Non-Invasive Diagnostics Of Ballistic Impact Damage," Proc. 25th Army Science Conf., Orlando, FL., Nov. 2006.

[8] J.M. Wells, "In-situ Fragment Analysis with X-ray Computed Tomography, XCT," ACERS, 31st Intn'l Conf. on Advanced Ceramics & Composites, 2007, In press

[9] J.M. Wells and R.M. Brannon, "Advances in XCT Diagnostics of Ballistic Impact Damage," **Dynamic Behavior of Materials**, Edited by N. Thadhani et al., TMS, *2007*, *In Press*

[10] N.L. Rupert, J.M. Wells, and W.J. Bruchey, "The Evolution and Application of Asymmetrical Image Filters for Quantitative XCT Analysis," 23rd Intn'l Symposium on Ballistics, Tarragona, Spain, ADPA v2, (2007) *In Press*

[11] J.M. Wells, N.L. Rupert, W.J. Bruchey, and D.A. Shockey, "XCT Diagnostic Evaluation of Ballistic Impact Damage in Confined Ceramic Targets", 23rd Intn'l Symp. on Ballistics, Tarragona, Spain, ADPA v2, (2007) *In Press*

[12] E. Straßburger: "Visualization of Impact Damage in Ceramics Using the Edge-On Impact Technique"; Int. Journal of Applied Ceramic Technology, Topical Focus: Ceramic Armor, American Ceramics Society, Vol. 1, Number 3, pp. 235-242, 2004

[13] E. Strassburger, P. Patel, J. W. McCauley, and D.W. Templeton, "High-Speed Photographic Study of Wave and Fracture Propagation in Fused Silica," Proceedings of 22nd IBS, ADPA, v2, 761-768 (2005).

[14] Straßburger, E., Patel, P., McCauley, J.W., Templeton, D.W., "Visualization of Wave Propagation and Impact Damage in a Polycrystalline Transparent Ceramic – AlON", Proc. 22nd Int. Symp. on Ballistics, 14-18 Nov., Vancouver, BC, Canada, 2005

[15] J.S. Steckenrider, W.A. Ellingson et al., "Evaluation of SiC Armor Tile Using Ultrasonic Techniques," Proc. of 30th Intn'l Conf. on Advanced Ceramics & Composites-Advances in Ceramic Armor, v27, 7, (2006), *In Press*.

[16] K. Thoma et al., "Real Time-Resolved Flash X-ray Cinematographic Investigation of Interface Defeat and Numerical Simulation Validation," 23rd Intn'l Symposium on Ballistics, Tarragona, Spain, ADPA, (2007) *In Press*

[17] P. Helberg, "High-Speed Three-Dimensional Tomographiic Imaging of Fragments and Precise Statistics From an Automated Analysis,", 23rd International Symposium on Ballistics, Tarragona, Spain, ADPA, (2007) *In Press*.

[18] C.E. Anderson, Jr., "A Review of Computational Ceramic Armor Modeling", Proceedings of the 30th International Conference on Advanced Ceramics and Composites," Ed. A. Wereszczak and E. Lara-Curzio, ACERS, J. Wiley, Paper 6-01, (2006).

[19] M. Wilkins, C. Honodel, and D. Sawle, "An approach to the study of light armor," UCRL-50284, Lawrence Livermore National Laboratory, Livermore, CA (1967).

[20] C. G. Fountzoulas, M.J. Normandia, J.C. LaSalvia, and B.A. Cheeseman, "Numerical Simulations of Silicon Carbide Tiles Impacted by Tungsten Carbide Spheres," Proceedings of 22nd IBS, ADPA, v2, 693-701 (2005).

[21] R.M. Brannon and J.M. Wells, "Validating Theories for Brittle Damage," **Dynamic Behavior of Materials,** Edited by N. Thadhani et al., TMS, *2007*, *In Press*

[22] Volume Graphics StudioMax v1.2.1, www.volumegraphics.com

ELASTIC PROPERTY MAPPING USING ULTRASONIC IMAGING

Raymond Brennan, Richard Haber, Dale Niesz, George Sigel
Rutgers University
607 Taylor Road
Piscataway, NJ, 08854-8065

James McCauley
US Army Research Laboratory
Aberdeen Proving Ground, MD, 21005-5066

ABSTRACT
Ultrasonic imaging is a nondestructive technique used for material inspection. Ultrasound, or acoustic, wave interactions with the material under inspection result in reflected signals from which time-of-flight (TOF) and signal amplitude values can be utilized to detect defects and inhomogeneities and to extract valuable information about elastic properties. With the use of an ultrasound scanning system, images can be generated based on changes in TOF and reflected signal amplitude. By measuring thickness and density variations over the material and collecting both longitudinal and shear TOF data, elastic property mapping can also be achieved. Elastic property mapping can provide valuable information for performance comparison and evaluation.

ULTRASOUND WAVE PROPAGATION
Ultrasonic, or acoustic, waves travel by exerting oscillating pressure on particles of a medium[1]. Since acoustic waves must displace a volume of material against the elastic constraints of its bonds in order to propagate, the ease of propagation of the wave is a function of density, which is related to the amount of material that must be moved, and a function of the elastic constraints, or how difficult it is to move the material. For this reason, the interactions between the acoustic waves and the medium can be used to provide density and elastic property measurements through ultrasonic testing of a material.

TIME-OF-FLIGHT, VELOCITY, AND ELASTIC PROPERTY MEASUREMENTS
The longitudinal and shear wave modes are critical for measuring elastic properties in solids. Their time of travel, or time-of-flight (TOF) through the specimen can be measured and used to determine the wave speed, or material velocity. The material velocities can be calculated by first measuring the longitudinal (TOF_l) and shear (TOF_s) time-of-flight values in addition to the thickness (t) of the specimen under evaluation. By utilizing the equations[2]:

$$c_l = 2t / TOF_l \tag{1}$$

$$c_s = 2t / TOF_s \tag{2}$$

the longitudinal (c_l) and shear (c_s) velocities can be calculated. The factor of two is used for a single transducer pulse-echo configuration in which the same transducer acts as a transmitter and receiver. This accounts for the round trip through the sample and back to the transducer. The density values can also be determined by using the equation[2]:

213

$$Z = \rho\, c_l \tag{3}$$

in which Z is the acoustic impedance of the material and ρ is the density. The acoustic impedance is a material property defined as the product of density and velocity. As a material property, the acoustic impedance should be the same for each specimen manufactured using the same materials, processing methods, and techniques. This property will vary for different materials and may vary for the same material as processing conditions are altered such as method of heat treatment methods and condition as well as addition of an additive or second phase.

Within an isotropic medium, the densities and longitudinal and shear velocities can be used to calculate elastic properties using the following equations[3,4]:

$$\nu = [1\text{-}2(c_s/c_l)]^2 \, / \, [2\text{-}2(c_s/c_l)]^2 \tag{4}$$

$$E = [(c_l)^2(\rho)(1\text{-}2\nu)(1+\nu)] \, / \, [(1\text{-}\nu)] \tag{5}$$

$$G = (c_s)^2(\rho) \tag{6}$$

$$K = E \, / \, [3(1\text{-}2\nu)] \tag{7}$$

where ν is Poisson's ratio, E is elastic modulus, G is shear modulus, and K is bulk modulus. By measuring the thickness of the specimen, the longitudinal TOF, and the shear TOF, a wide variety of important properties can be calculated.

POINT ANALYSIS

The sample under examination was a four by four inch sintered silicon carbide (SiC) plate. The sample was evaluated at nine separate points, as shown in Figure 1. The values at each of the nine points are shown in Table I. The thickness at each point was first measured using an electronic caliper, and the average thickness of the sample was found to be 7.68 mm. The sample density was also measured using Archimedes method, and found to be 3.219 g/cm³. Ultrasound point analysis was conducted using a 50 MHz longitudinal contact transducer and a 25 MHz shear contact transducer to get an idea of TOF, velocity, and elastic property values across the sample. By using the longitudinal wave transducer to measure the difference in microseconds between the top surface reflected signal and the bottom surface reflected signal in the specimen, the longitudinal TOF was determined at each point, as shown in Table I. The top and bottom surface reflected signals used to calculate longitudinal TOF are shown in Figure 2. The average longitudinal TOF was found to be 1.348 µs. One noticeable trend was that the longitudinal TOF values were lower at points 1, 4, and 7, with an average of 1.327 µs, as compared to the rest of the sample which had an average longitudinal TOF value of 1.359 µs. Since points 1, 4, and 7 represent the left side of the sample, it was important to keep this difference in mind during further evaluation.

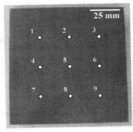

Figure 1. Nine points used for ultrasound point analysis of sintered SiC plate.

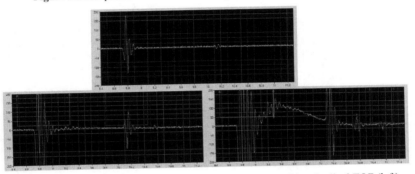

Figure 2. Amplitude scans of reflected signals (top) and gated longitudinal TOF (left) and shear TOF (right) peaks.

By using the shear wave transducer to measure the difference in microseconds between the top surface reflected signal and the reflected shear signal, which is approximately 1.5 to 1.7 times greater than the longitudinal TOF[4], the shear TOF was also determined, as shown in Table 1. The top surface reflected signal and mode-converted shear signal, shown in Figure 2, were used to calculate shear TOF. The average shear TOF was found to be 2.117 µs. Just as with the longitudinal TOF values, the shear TOF values were lower at points 1, 4, and 7, with an average of 2.083 µs, as compared to the rest of the sample which had an average longitudinal TOF value of 2.134 µs. By using the longitudinal and shear velocity equations for a single transducer pulse echo configuration, the average longitudinal velocity was found to be 11,400 m/s and the average shear velocity was 7,220 m/s. Using the elastic property equations described earlier, the average Poisson's ratio (ν), elastic modulus (E), shear modulus (G), and bulk modulus (K) values were calculated as 0.16, 390 GPa, 170 GPa, and 190 GPa, respectively. While these values were very similar across each of the nine points, the E values of 410 GPa at point 1 and 400 GPa at point 4, the G value of 180 GPa at point 1, and the K values of 200 GPa at points 1, 4, and 7 were all above the averages. These values also occurred along the left side of the sample.

ULTRASONIC IMAGING

Ultrasonic imaging can be performed by immersing a test specimen in water and mechanically translating an ultrasonic transducer over the specimen to collect reflected signals caused by Z differences between inhomogeneities and the bulk. The reflected signals from the water/top specimen surface and bottom specimen surface/water can be used to measure longitudinal and shear TOF values. By using the proper longitudinal ultrasound transducer for immersion testing and C-scan imaging, both longitudinal and shear waves reflections can be determined due to mode conversion. In general, vibrational waves may change their mode of vibration or be subject to mode conversion at material interfaces[2]. Mode conversion occurs when the longitudinal waves in the water are converted to shear waves in the material due to refraction of the waves at the interface[2].

ELASTIC PROPERTY MAPPING

While time-of-flight (TOF) and reflected signal amplitude maps are common forms of C-scan imaging that have been well established in ultrasound, the next generation of image mapping is focusing on visually depicting variations in longitudinal velocity, shear velocity, Poisson's ratio, elastic, shear, and bulk modulus. Through the use of a specialized broadband transducer capable of achieving sharp, high intensity reflections at material boundaries, signals were generated with sufficient signal-to-noise ratio for evaluation of both longitudinal and shear wave reflected signals. The additional equipment, parameters, and procedures utilized for performing ultrasound C-scan imaging and collecting reflected ultrasound signals are described in the referenced articles[5,6].

By using a C-scan imaging system to collect enough data points for effective mapping, over 200,000 TOF values exhibiting the changes in longitudinal and shear TOF were measured and mapped, as shown in Figure 3. While the top and bottom sample surfaces demonstrated high intensity reflected signals for measuring the longitudinal TOF, the shear peak was more highly attenuated than the longitudinal peak, and the capabilities of the broadband transducer for detecting the mode-converted shear signal from the specimen were necessary.

Figure 3. Longitudinal TOF (left) and shear TOF (right) C-scan image maps in which the darker colors represent lower material velocity values at a constant thickness. Since the thickness of the test specimen is nonuniform, it is evaluated using a separate image map.

In addition to these two parameters, thickness variations and density variations over the scanned regions were also critical for obtaining the necessary parameters for measuring the elastic properties. By running a low frequency TOF ultrasound scan at 1 MHz, as shown in Figure 4, a map of the changes in thickness was collected. This was possible due to the fact that the specimen was a high density plate with relatively minor material inhomogeneities. If there were anomalous defects or regions with drastic density changes or inhomogeneities, this would have affected the strict evaluation of thickness and lower frequencies would have been used.

Figure 4. Low frequency longitudinal TOF scan for assessing thickness variations in which darker colors represent slightly thicker regions.

The longitudinal TOF, shear TOF, and thickness variation plots were used to map longitudinal and shear velocities using the aforementioned equations. The velocity maps are shown in Figure 5. The density variations were obtained by using the collected longitudinal velocity values in addition to the material property of acoustic impedance, which was assumed to be constant for the same type of sintered SiC material. By taking an average of eight sintered SiC samples manufactured under the same conditions, the acoustic impedance for these tiles was found to be 36.9×10^5 g/cm^{2}s. By using this material property as the Z value in addition to measured changes in c_l, densities changes were calculated over the scanned area. With a full set of data including c_l, c_s, t, and ρ values at each x/y position on the map and the gravitational constant value assumed to be 9.81 m/s^2, elastic properties were calculated at each point using the aforementioned equations. These values were mapped at each x/y position and color scaled based on minimum and maximum values to obtain image maps of each respective elastic property, as shown in Figures 6 and 7.

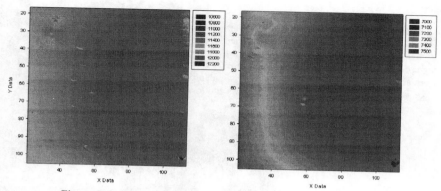

Figure 5. Longitudinal (left) and shear (right) velocity image maps.

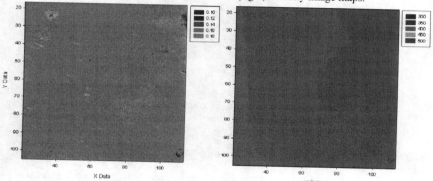

Figure 6. Poisson's ratio (left) and elastic modulus (right) image maps.

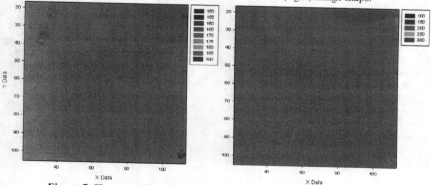

Figure 7. Shear modulus (left) and bulk modulus (right) image maps.

From the image maps for the sintered SiC plate, regional variations and changes in elastic properties from isolated defects were found. In the TOF images, multiple isolated areas were located that had a different average TOF as compared to the rest of the bulk. The two largest areas in which this occurred were in the top left corner of the sample, which had a lower TOF value, and the bottom right corner of the sample, which had a higher TOF value than the average. The two distinct regions will be compared for each scan. Besides these isolated features, the TOF images showed a slight gradient of lower to higher TOF values from the left to the right side of the sample. These were the same trends found during point analysis of the sample, with TOF values on the left side averaging 1.327 μs for longitudinal and 2.083 μs for shear as compared to 1.359 μs for longitudinal and 2.134 μs for shear over the rest of the sample area. By applying the thickness correction factor, these patterns were accounted for so that any remaining differences were not due to changes in thickness, but changes in the bulk properties.

The same trends found in the TOF maps were analyzed using the velocity and elastic property maps. In the velocity maps, the low TOF area in the top left corner corresponded to the area of highest longitudinal and shear velocity, ranging between 11,800 to 12,000 m/s for longitudinal velocity as compared to the average velocity of 11,400 m/s, and between 7,400 and 7,500 m/s for shear velocity as compared to the average velocity of 7,220 m/s. The high TOF area in the bottom right corner corresponded to the area of lowest longitudinal and shear velocity, ranging between 10,800 and 11,000 m/s for longitudinal velocity as compared to the average velocity of 11,400 m/s, and between 6,900 m/s and 7,000 m/s for shear velocity as compared to the average velocity of 7,220 m/s. Even after accounting for the change in thickness, the overall velocity trend for the sample decreased from left to right, which correlated to the point analysis data trends. For points 1, 4, and 7 on the left side of the sample, the average longitudinal velocity was 11,560 m/s, which decreased to an average of 11,340 m/s for points 2, 5, and 8 down the center of the sample, and further decreased to an average of 11,300 m/s for points 3, 6, and 9 on the right side of the sample. Similar trends were found for the shear velocity map.

Evaluation of the elastic property maps showed a more narrow range of variation, but evidence of similar trends was found as compared to the longitudinal and shear TOF and velocity maps. In the Poisson's ratio point analysis data, the values at all nine points were found to be 0.16. The majority of the v map matches this value as well, though there are some deviating areas. There are several areas of higher v with values above 0.18 scattered throughout the sample, including the isolated features in the top left and bottom right corners. In the center of the bottom right feature, v dropped to about 0.14. In the elastic modulus map, there are three distinct regions from left to right that represented E values ranging from 400 to 410 GPa on the left edge, 390 to 400 GPa through most of the center, and 380 to 390 GPa on the right side. While the isolated feature in the top left corner was not detectable, the one on the bottom right showed an E value ranging between 370 and 380 GPa. The shear modulus maps showed more of a gradient from left to right, with higher G values around 180 GPa on the left side of the sample and lower G values of 160 GPa on the right side of the sample. The isolated feature in the bottom right had a G value of approximately 150 GPa. The bulk modulus map displayed a very narrow range, with the majority of K values falling between 190 and 200 GPa over the sample area. As expected, the bottom right feature had a lower K value of approximately 150 GPa, but there were two additional regions with K values ranging between 160 and 170 GPa above the bottom right feature and in the top right corner. Overall, the elastic property trends were consistent with point analysis measurments, with values decreasing from left to right. Another important trend was that the feature at the bottom right of the sample consistently had the lowest elastic properties. This isolated region would be red-flagged as the weakest point in the material.

The image map data was collected and summarized in Table II in terms of the mean, standard deviation, standard error, and 95% and 99% confidence interval for each data set, representing thickness, density, longitudinal TOF and velocity, shear TOF and velocity, Poisson's ratio, elastic modulus, shear modulus, and bulk modulus, respectively.

CONCLUSIONS

Ultrasound C-scan imaging was utilized for collection of both longitudinal and shear TOF image maps. After correcting the data for thickness variations and calculating density changes, equations were used to obtain longitudinal velocity, shear velocity, Poisson's ratio, elastic modulus, shear modulus, and bulk modulus values that were converted to image maps. The mapping results and trends were consistent with ultrasound point analysis data. While further investigation will be conducted on specific local regions, the initial collected image maps appear to be valid assessments of variations in material velocities and elastic properties obtained using ultrasound imaging.

ACKNOWLEDGEMENTS

The authors would like to thank the U.S. Army Research Laboratory's Material Center of Excellence - Lightweight Materials for Vehicle Protection Program, Cooperative Agreement No. W911NF-06-2-0007 and the Ceramic and Composite Materials Program an NSF I/UCRC, Agreement No. EEC-0436504, for their support.

REFERENCES

[1]M.C. Bhardwaj, "Evolution, Practical Concepts and Examples of Ultrasonic NDC", *Ceramic Monographs – Handbook of Ceramics*, **41**, 1-7 (1992).

[2]P.E. Mix, *Introduction to Nondestructive Testing*, John Wiley & Sons, 104-153 (1987).

[3]Albert Brown; Rationale and Summary of Methods for Determining Ultrasonic Properties of Materials at Lawrence Livermore National Laboratory (http://www.llnl.gov/tid/lof/documents/pdf/225771.pdf)

[4]N. Kulkarni, B. Moudgil, M. Bhardwaj, "Ultrasonic Characterization of Green and Sintered Ceramics: I, Time Domain", *American Ceramic Society Bulletin*, **73** [6] 146-153 (1994).

[5]R.E. Brennan, R. Haber, D. Niesz, J. McCauley, and M. Bhardwaj, "Non-Destructive Evaluation (NDE) of Ceramic Armor: Fundamentals"; *Advances in Ceramic Armor*, **26** [7] 223-230 (2005).

[6]R.E. Brennan, R. Haber, D. Niesz, and J. McCauley, "Non-Destructive Evaluation (NDE) of Ceramic Armor: Testing", *Advances in Ceramic Armor*, **26** [7] 231-238 (2005).

Table I. Point analysis and average value results.

#	ρ (g/cc)	t (mm)	TOF_l (μs)	TOF_s (μs)	c_l (m/s)	c_s (m/s)	ν	E (GPa)	G (GPa)	K (GPa)
1	3.219	7.69	1.328	2.085	11,580	7,380	0.16	410	180	200
2	3.219	7.70	1.355	2.127	11,370	7,240	0.16	390	170	190
3	3.219	7.72	1.367	2.146	11,290	7,190	0.16	390	170	190
4	3.219	7.63	1.326	2.082	11,510	7,330	0.16	400	170	200
5	3.219	7.70	1.357	2.130	11,350	7,230	0.16	390	170	190
6	3.219	7.68	1.362	2.138	11,280	7,180	0.16	390	170	190
7	3.219	7.68	1.326	2.082	11,580	7,040	0.16	390	170	200
8	3.219	7.67	1.357	2.130	11,300	7,200	0.16	390	170	190
9	3.219	7.69	1.358	2.132	11,330	7,210	0.16	390	170	190
Avg	**3.219**	**7.68**	**1.348**	**2.117**	**11,400**	**7,220**	**0.16**	**390**	**170**	**190**

Table II. Quantitative values from each set of data points.

Parameter	t (mm)	ρ (g/cc)	TOF_l (μs)	TOF_s (μs)	c_l (m/s)	c_s (m/s)	ν	E (GPa)	G (GPa)	K (GPa)
Mean	7.68	3.221	1.348	2.117	11,400	7,260	0.16	393	170	192
Std Dev	0.017	0.035	0.017	0.027	123.57	76.20	0.004	4.14	1.96	3.08
Std Error	3.8E-5	7.6E-5	3.8E-5	5.9E-5	0.272	0.168	9.3E-6	9.1E-3	4.3E-3	6.7E-3
95% Conf	7.4E-5	1.5E-4	7.4E-5	1.2E-4	0.532	0.328	1.8E-5	0.018	0.008	0.013
99% Conf	9.7E-5	2.0E-4	9.8E-5	1.5E-4	0.700	0.432	2.4E-5	0.023	0.011	0.017

Author Index

Author Index